The Experimental and Historical Foundations of Electricity

Andre Koch Torres Assis

Apeiron
Montreal

Published by C. Roy Keys Inc.
4405, rue St-Dominique
Montreal, Quebec H2W 2B2 Canada
http://redshift.vif.com

First Published 2010

Library and Archives Canada Cataloguing in Publication

Assis, André Koch Torres, 1962-
The experimental and historical foundations of electricity / André Koch Torres Assis.

Includes bibliographical references.
ISBN 978-0-9864926-3-1

1. Electricity--Experiments. 2. Electricity--History. I. Title.

QC533.A88 2010 537'.078 C2010-901025-6

Front cover: The experiment by Stephen Gray (1666-1736) with the suspended boy (1731) as represented in Doppelmayr's book, *Neu-entdeckte Phaenomena von bewunderswürdigen Würkungen der Natur*, Nurenburg, 1774. A boy is suspended by insulating lines. A rubbed glass tube is brought near his legs. The hands and face of the boy attract light bodies.

Back cover: Photos of instruments described in this book. A metal versorium. A Du Fay versorium made of plastic with the tip of one of its legs wrapped in aluminum foil. An electric pendulum with a paper disk attached to a silk thread tied to a plastic straw. An electrified electroscope with its raised strip made of tissue paper. The thin cardboard of the electroscope is attached to a plastic straw.

Contents

Presentation and Acknowledgments

In the early 1990's I discovered the work of Norberto Cardoso Ferreira, of the Institute of Physics at the University of São Paulo, USP, Brazil. One of his research interests was to experimentally demonstrate the most important aspects of electricity utilizing very simple and easily available materials. I had the opportunity to visit him at USP in 1993. During this visit he gave me a small set of experimental materials made of thin cardboard, plastic straws, tissue paper, paper fasteners, etc. He showed me how to perform the main experiments and also showed me his book *Plus et Moins: Les Charges Électriques.*[1] I became fascinated with what I learned, realizing how it was possible to experimentally envision very profound physical phenomena dealing with easily found materials. I kept this material as a treasure for 10 years, but neither used nor developed it during this period. I am extremely thankful to Norberto Ferreira for what I learned from him. Recently I discovered other works by Ferreira, as always extremely rich and creative.[2] I also learned during discussions with his students, like Rui Manoel de Bastos Vieira and Emerson Izidoro dos Santos.

In 2005 I met Alberto Gaspar and discovered his book *Experiências de Ciências para o Ensino Fundamental.*[3] I also learned a great deal from his book and other of his works.[4]

Between 2004 and 2007 I taught classes to high school science teachers in the *Teia do Saber* project of the Secretary of Education of the State of São Paulo, in Brazil. It was a great privilege to be invited to participate in this project. The support I received from the Secretary of Education and from the Coordinating Group of Educational Projects of the University of Campinas,

---

[1] [FM91].

[2] [Fer78], [Fera], [Ferb], [Ferc], [Ferd], [Fer06], [Fer01c], [Fer01d], [Fer01b], and [Fer01a].

[3] [Gas03].

[4] [Gas91] and [Gas96].

GGPE—UNICAMP, as well as the rich contacts with high school science teachers who took our classes, were extremely productive and stimulating for me. I also profited greatly from many exchanges of ideas with professors at the University of Campinas who participated in this project. As part of my activities, I decided to teach the high school science teachers what I had learned with Norberto Ferreira. As a result, I returned to the experiments with the further motive of writing this book, in order to share all this fascinating material with a wider audience.

The inspiration for the majority of the experiments described in this book was taken from the original works of the scientists discussed here, and from the books and papers of Norberto Ferreira and Alberto Gaspar. Since 2004 I have discovered other printed works and interesting websites which have been extremely helpful to my apprenticeship in this area—such as the site *Feira de Ciências*, organized by Luiz Ferraz Netto.[5]

John L. Heilbron suggested relevant improvements in the first version of this book. His great work, *Electricity in the 17th and 18th Centuries: A Study in Early Modern Physics*,[6] was our main source of historical information related to electrostatics. Many important suggestions to improve an earlier version of this work have also been given by Sérgio Luiz Bragatto Boss, John Eichler, Steve Hutcheon, Fabio Miguel de Matos Ravanelli, and Bertrand Wolff.

The Figures for this book were prepared by Daniel Robson Pinto, through a fellowship awarded by SAE/UNICAMP, which we thank for this support. Daniel helped also to obtain old figures and references.

I would also like to thank several other people for suggestions, discussions, and references: Christine Blondel, Paolo Brenni, João José Caluzi, Juliano Camillo, Hugo Bonette de Carvalho, João Paulo Martins de Castro Chaib, Asit Choudhuri, Roberto Clemente, Junichiro Fukai, Hans Gaab, Robert Harry van Gent, Harald Goldbeck-Löwe, Jürgen Gottschalk, Peter Heering, Elizabeth Ihrig, John Jenkins, Siegfried Kett, Ellen Kuhfeld, Wolfgang Lange, Lin Liu, José Joaquín Lunazzi, Ceno Pietro Magnaghi, Eduardo Meirelles, Mahmoud Melehy, Dennis Nawrath, Marcos Cesar Danhoni Neves, Horst Nowacki, Martin Panusch, José Rafael Boesso Perez, Karin Reich, Edson Eduardo Reinehr, Ricardo Rodrigues, Waldyr Alves Rodrigues Jr., Torsten Rüting, Dirceu Tadeu do Espírito Santo, Wayne M. Saslow, Fernando Lang da Silveira, Moacir Pereira de Souza Filho, Christian Ucke, Alvaro Vannucci, Geraldo Magela Severino Vasconcelos, Greg Volk, Karl-Heinrich Wiederkehr, Bernd Wolfram, and Gudrun Wolfschmidt.

I wish to thank the Institutes of Physics and Mathematics, the Coordinating Group of Educational Projects, and the Fundo de Apoio ao Ensino, à Pesquisa e à Extensão—FAEPEX of the University of Campinas—UNICAMP, which gave the necessary support for undertaking this work. I thank also the Institute for the History of Natural Sciences of Hamburg University and the Alexander von Humboldt Foundation of Germany for a Research Fellowship in the year 2009

[5] [Net].
[6] [Hei99].

during which we collected considerable bibliographic material for this book.

Roy Keys, the Editor of Apeiron, has been a supporter for many years. Without his encouragement some of my books might not have been published. He made an excellent editorial work for this book.

Andre Koch Torres Assis
Institute of Physics
University of Campinas—UNICAMP
13083-970 Campinas - SP, Brazil
E-mail: assis@ifi.unicamp.br
Homepage: www.ifi.unicamp.br/~assis

Chapter 1

Introduction

One of the goals of this book is to present the basic phenomena of electricity through simple experiments performed with low cost materials. We describe experiments about attraction and repulsion; we show how to charge bodies by friction/contact/induction; we analyze the different properties of conductors and insulators, etc. Moreover, we show how theoretical concepts are formed and modified in this process, just as happens with the fundamental laws describing these phenomena.

We then illustrate how more complex phenomena can be understood and clarified in terms of the earlier elementary experiments. Some playful and curious experiments are also presented. They are designed to stimulate creativity and critical thinking. Some also seek to relate daily phenomena to the basic laws of physics.

The emphasis is placed on experimental activities. Beginning with the experiments, we formulate the definitions, concepts, postulates, principles, and laws that describe the phenomena. The materials utilized are very simple, easily available at home or in stores, and all inexpensive. Nevertheless, with them we perform precise experiments and build very sensitive scientific instruments. The reader thus need not depend on a school or research laboratory, as he will build his or her own equipment and perform all the measurements. To help achieve this goal, we present several different assemblies for each instrument and more than one way of performing the measurements.

Another important motivation we had in mind before writing this book was to offer teachers and students the main tools to achieve scientific autonomy. To do this, we quote sections of the most important works by the scientists who made great and fundamental discoveries in electricity. We also show how to perform experiments illustrating their findings utilizing low cost instrumentation. In this way we hope the readers will attain a scientific independence in several respects: how to build instruments, how to perform measurements, how to formulate concepts and theories to clarify or explain their findings, etc.

If the experiments presented here are performed in the classroom, each student should perform the activities and build his own equipment (electroscope, versorium,[1] electric pendulum). The students should take all this personal material home. This procedure is much more rewarding than when the experiments are merely demonstrated by the teacher where students normally do not put their hands to the plough. We believe that hands-on science is one of the most effective learning techniques.

Beyond the experimental part, this book is also rich in historical information, which provides the context in which some phenomena and laws arose. It also gives different interpretations of these observational facts. Special care is taken in formulating and stating the concepts and physical principles covered. We try to be careful with the words used, making every effort to distinguish clearly between definitions, postulates, and experimental results. We also distinguish the description from the explanation of a phenomenon. The aim is to illustrate the human and sociological elements embodied in the formulation of physical laws. We do not follow the historical sequence of the discoveries. However, whenever possible we describe the context in which each phenomenon was discovered, and also mention the principal scientists involved in the discovery. The main historical information presented here was taken from the original works quoted in the text and also from the excellent books by Heilbron.[2] Our goal is not to present the different explanations and theoretical models that have been proposed through the centuries to account for electrical phenomena. Heilbron's books are some of the best sources for anyone interested in these last aspects.

In order to keep this book within a reasonable size, we have chosen a few specific topics to be discussed in greater detail. In a future book we hope to deal with other important aspects of electricity following a similar procedure. At that point we will deal with sparks and discharges, the power of points, the electric wind, Volta's electrophorus, Leyden's jar, Faraday's cage, Gray and the preservation of electrical charges (the longevity of the electrification of objects, or how to store electricity for a long time), Ohm's law, contact/rolling/separation electrifications, charge generators, electrets, capacitance and charge distribution between conductors, atmospheric electricity, Lichtenberg figures, etc.

In the present book we show that many fundamental questions of science can be explored by means of experiments performed with very simple materials which nevertheless have great historical or conceptual importance. Throughout the text we show that some of the great scientists of history dealt with these phenomena, which nowadays seem so simple and trivial, but which actually still conceal very deep mysteries.

This book is written for teachers and students of physics, mathematics, science, and engineering. It is not a book of experiments for children. It can be used in high schools or universities, depending on the level at which each phenomenon or law is analyzed. It has experimental and theoretical material which can be applied at all levels of teaching. Each teacher may choose the material presented here and adapt it to his teaching environment. It can also be utilized in history and philosophy of science courses. Parts of this work can

[1] See Chapter 3.

[2] [Hei79], [Hei82], and [Hei99].

even be utilized at a post-graduate level or for further scientific research.

The best way to follow this book is to perform the majority of the experiments described here as you read, rather than simply read the contents. The preferred approach is to try to repeat, perfect, and modify what is proposed here. Although physics has many different aspects—philosophical, theoretical, and mathematical—it is essentially an experimental science. It is the coming together of all these aspects that makes it so fascinating. We hope the reader will feel the same pleasure in performing these experiments as we had in devising them.

I would like to receive a feedback from readers who have tried to reproduce and develop the experiments described here, or attempted to apply them at their schools and universities. I myself, particularly, would have greatly enjoyed learning physics in this way. That is, instead of learning several formulas by heart and spending most of my time solving mathematical exercises, I would prefer to learn physics in the manner shown here, by having the opportunity to build instruments and perform various experiments, learning in practice how important phenomena were first discovered and interpreted, and reproducing most of these effects with simple materials myself. It would also be very interesting to explore different models and theoretical concepts in order to explain these phenomena. This book is our contribution to improving the teaching of physics, in a manner similar to what we did with the concept of the center of gravity and the law of the lever.[3] We hope that science can thus be presented in a more palpable way, rich in historical context, such that the creativity and critical mind of the students can be stimulated.

I would be happy if this book were translated to other languages. It would be great if teachers of physics might indicate this material to their colleagues and students. I also hope it will motivate others to try something similar in other areas of science, utilizing experiments performed with accessible materials combined with historical information related to the subject.

Before beginning these experiments we should call the reader's attention to a few important points. Normally the experiments work well on cold and dry days. On humid, hot days, or when it is raining, many effects described here may not be seen, or the phenomena may be of low intensity, making it difficult to observe them. On a number of occasions in the book we quote generic names of substances like plastic, glass, wood, or rubber. But it should be borne in mind that there are in reality many varieties of plastic, glass, wood, rubber, or any other substance. These varieties are intrinsically different from one another due to their composition, fabrication process, age, etc. Therefore, when a specific effect is not observed with a certain substance (with a specific type of plastic, for instance), one should try the same experiment with another analogous substance to see what happens.

A Portuguese version of this book is being published under the title *Os Fundamentos Experimentais e Históricos da Eletricidade.*

The quotations in English are taken from the original works or from quoted

[3] [Ass08a] and [Ass08b].

English translations; otherwise they were done by ourselves. The expressions between square brackets in the middle of some quotations are ours, intended to clarify the meaning of certain sentences.

Chapter 2

Electrification by Friction

2.1 The Beginning of the Study of Electricity

Experiment 2.1

In the first experiment we cut several small pieces of paper and place them on a table. We take a plastic straw and bring it close to the pieces of paper, taking care not to touch the paper. Nothing happens to the pieces of paper (Figure 2.1).

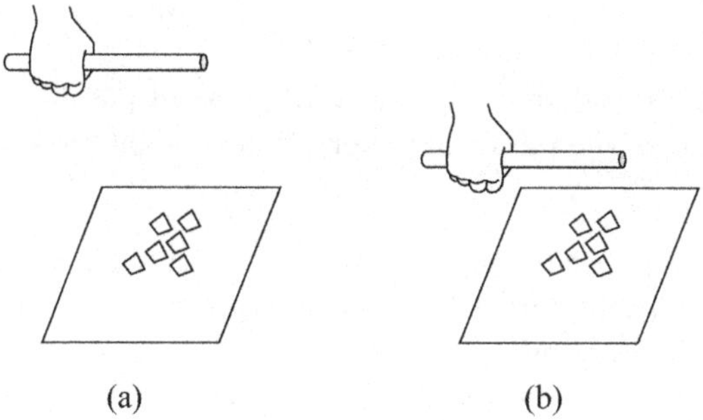

Figure 2.1: (a) Plastic straw far away from pieces of paper. (b) When the plastic straw is moved near the pieces of paper, nothing happens to them.

We now rub the straw in our hair or in a sheet of paper, moving it briskly up and down. We represent the region of the straw which has been rubbed by the letter F, taken from the word *friction* (Figure 2.2).

We then bring the rubbed straw near the small pieces of paper, once more without touching them, only coming very close. We observe that beyond a certain distance they jump to the rubbed straw and remain attached to it (Figure 2.3). As we move the straw away from the table, the pieces of paper remain attached to the straw.

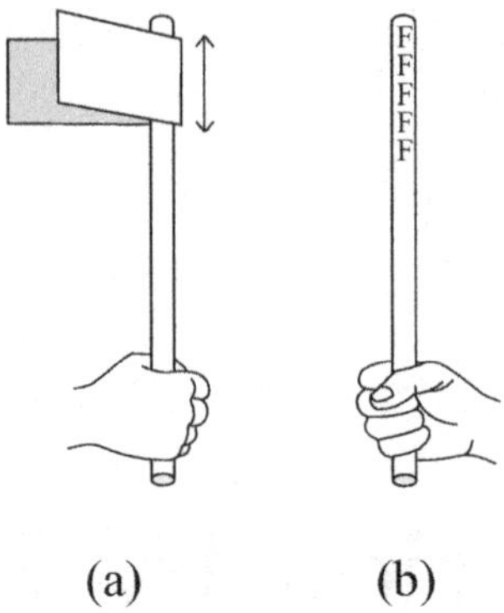

Figure 2.2: (a) Plastic straw rubbed by paper. (b) The letter F represents the rubbed region of the straw.

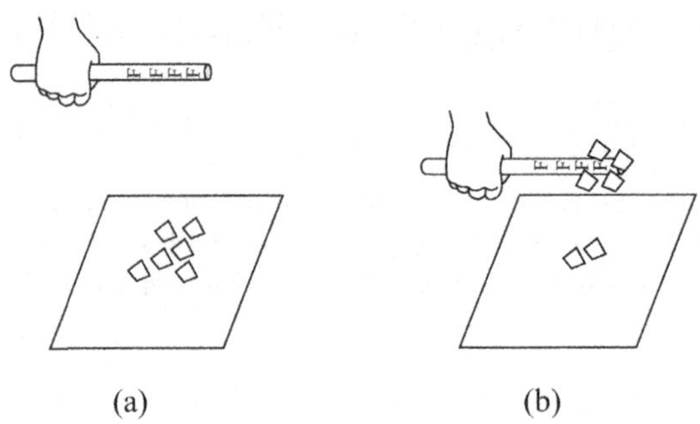

Figure 2.3: (a) A rubbed straw far away from small pieces of paper. (b) The rubbed straw attracts the pieces of paper when brought close to them.

Not all pieces of paper remain attached to the rubbed straw. Some of them touch the straw and fall. Others are reflected back to the table. This will be discussed in Sections 4.4 and 4.8.

An analogous experiment can be done with the plastic body of a pen, a plastic ruler, or a plastic comb. In order to avoid complicated phenomena or unexpected results, these items should be made only of plastic, without metal parts, etc. Nothing happens to the small pieces of paper when we move these objects near them, provided these bodies were not previously rubbed. We now rub any of these bodies with hair or a sheet of paper. We then move the rubbed plastic near the pieces of paper, without touching them. They are again attracted to the plastic, remaining attached to it. Each person should find an appropriate plastic material which, when rubbed, easily attracts small pieces of paper. Normally we will mention straws, but we can also employ rulers or combs, depending on what is available or what creates a larger effect.

Definitions: Usually we say that plastic which has not been rubbed and which does not attract small pieces of paper is *electrically neutral* or, simply, *neutral*. When it has been rubbed, we say that the plastic has *acquired an electrical charge*, become *electrified*, *electrically charged* or, simply, *charged*. The process is called *charge generation by friction*, *charge generation by rubbing*, *frictional electrification*, *triboelectrification*, *electrification by rubbing*, or *electrification by friction*. This attraction is sometimes referred to as an *electric attraction*, or as an *electrostatic attraction*.

In this and other experiments in this book we will refer to friction between a plastic body and hair (or between the plastic and a sheet of paper, or between the plastic and a tissue). In order for the experiments to succeed, it is a good practice to surround the plastic body with a separate second piece of paper and to hold it firmly in our hands. We then move the body and paper quickly in opposite directions as we press them together. Usually it helps to perform this motion for each body in a single direction, instead of making a to-and-fro motion. For instance, we can move the plastic toward our body and the paper away from us. It also helps to repeat this procedure more than once before beginning any experiment, as this enhances the effects to be observed. From time to time it is convenient to replace the materials to observe how the effect varies with different materials. Practice will dictate the best materials and procedures for a successful experiment.

2.2 The Amber Effect

Most of us have performed an experiment like this first one as a child or teenager. And it was with an experiment analogous to this one that the whole science of electricity was born! Since at least Plato (circa 428-348 B.C.) it has been known that rubbed amber attracts light objects placed near it. The oldest mention of this fact, sometimes known as the "amber effect," appears in his dialogue, *Timaeus*:[1]

> Let us once more consider the phenomena of respiration, and enquire into the causes which have made it what it is. They are as follows:—Seeing that there is no such thing as a vacuum into which any of those things which are moved can enter, and the breath is carried from us into the external air, the next point is, as will be clear to every one, that it does not go into a vacant space, but pushes its neighbour out of its place, and that which is thrust out in turn drives out its neighbour; and in this way everything of necessity at last comes round to that place from whence the breath came forth, and enters in there, and following the breath, fills up the vacant space; and this goes on like the rotation of a wheel, because there can be no such thing as a vacuum. [...] Moreover, as to the flowing of water, the fall of the thunderbolt, and the marvels that are observed about the attraction of amber and the Heraclean stones,—in none of these

[1] [Pla52b, Sections 79 to 80, pp. 470-471].

cases is there any attraction; but he who investigates rightly, will find that such wonderful phenomena are attributable to the combination of certain conditions—the non-existence of a vacuum, the fact that objects push one another round, and that they change places, passing severally into their proper positions as they are divided or combined.

Such as we have seen, is the nature and such are the causes of respiration—the subject in which this discussion originated.

He does not mention who discovered this fact, but from his casual description it seems that the amber effect was known to his readers. He connects the amber effect with that of the Heraclean stone, or natural magnet. Plato rejects the idea that there is a real attraction between the rubbed amber and the light objects nearby. All of these phenomena are explained on the same principles as in respiration, namely, the non existence of a vacuum.

Amber is a hard yellowish to brownish translucent resin,[2] sometimes used for jewelry. Since the 19th century it has been known that it is the fossil resin of pine trees which have probably been dead for many millions of years.[3] Some stores sell amber with fossilized insects inside, like ants, fleas, or spiders. In Figure 2.4 we see two pieces of amber.

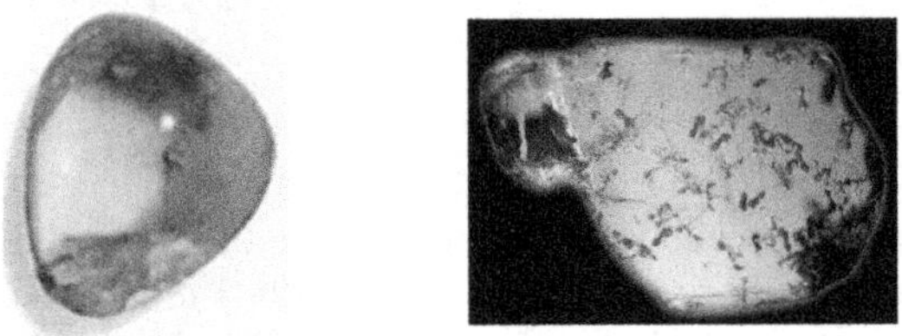

Figure 2.4: Pieces of amber.

Aristotle (384-322 B.C.), in his work *Meteorology*, presented evidence that the amber occurred originally in liquid form and later solidified:[4]

> Of solid bodies those that have been solidified by cold are of water, e.g. ice, snow, hail, hoar-frost. [...] Of these bodies those from which all the moisture has gone are all of them of earth, like pottery or amber. (For amber, also, and the bodies called 'tears' are formed by refrigeration, like myrrh, frankincense, gum. Amber, too, appears to belong to this class of things: the animals enclosed in it show that it is formed by solidification. The heat is driven out of it by the cold of the river and causes the moisture to evaporate with it, as in the case of honey when it has been heated and is immersed in water.)

[2]See Appendix A.
[3][RR53].
[4][Ari52b, p. 492].

According to some modern authors, the amber experiment was performed for the first time by Thales of Miletus, who lived from around 625 to 546 B.C. Plato names him first on his list of the seven sages of ancient Greece in his dialogue *Protagoras*.[5] But he does not attribute the amber effect to him. Thales is considered by Aristotle and by several ancient writers to be the first natural philosopher, or as the first physicist. In his book *Metaphysics* (A 3, 983 b 6), Aristotle wrote the following about him (our emphasis in italics):[6]

> Of the first philosophers, then, most thought the principles which were of the nature of matter were the only principles of all things. That of which all things that are consist, the first from which they come to be, the last into which they are resolved (the substance remaining, but changing in its modifications), this they say is the element and this the principle of things, and therefore they think nothing is either generated or destroyed, since this sort of entity is always conserved, as we say Socrates neither comes to be absolutely when he comes to be beautiful or musical, nor ceases to be when he loses these characteristics, because the substratum, Socrates himself remains. Just so they say nothing else comes to be or ceases to be; for there must be some entity—either one or more than one—from which all other things come to be, it being conserved.
>
> Yet they do not all agree as to the number and the nature of these principles. *Thales, the founder of this type of philosophy*, says the principle is water (for which reason he declared that the earth rests on water), getting the notion perhaps from seeing that the nutriment of all things is moist, and that heat itself is generated from the moist and kept alive by it (and that from which they come to be is a principle of all things). He got his notion from this fact, and from the fact that the seeds of all things have a moist nature, and that water is the origin of the nature of moist things.

However, none of Thales's works has come down to us. The origin of all modern claims relating Thales to the amber experiment is in the writings of Diogenes Laertius, who lived around the third century A.D. and was a biographer of Greek philosophers. His most important work is called the *Lives of Eminent Philosophers*, in 10 volumes. He said the following about Thales:[7]

> [...] He was the first to give the last day of the month the name of Thirtieth, and the first, some say, to discuss physical problems.
>
> Aristotle and Hippias affirm that, arguing from the magnet and from amber, he attributed a soul or life even to inanimate objects.

[5] [Pla52a, pp. 54-55].
[6] [Ari52a, Book 1, Chapt. 3, p. 501-502].
[7] [Lae91, p. 25].

Another translation of this passage is the following:[8] "Aristotle and Hippias say that, judging by the behaviour of the lodestone and amber, he also attributed souls to lifeless things."

Normally a soul is attributed to something which is alive or which can move by its own will. Sometimes a soul is also attributed to something which can grow, like a man, a plant, or an animal. Those are the living or animate bodies. The inanimate bodies, or bodies without a soul, are those not endowed with life. Although magnets and amber neither grow nor move by themselves, they have the property of generating motion of nearby objects (like a magnet attracting iron or being attracted by iron, or a rubbed amber attracting chaff). Thales may have attributed a soul to a magnet or amber due to these properties.

Despite these statements by Diogenes Laertius, it is doubtful that Thales was really the first to perform the amber experiment.[9] He is considered by later authors to be the initiator of many things in physics and in mathematics, which casts some doubt on the reality of all these achievements. As regards the above statement, it is difficult to verify Laertius's sources. Hippias's writings are lost to us. As for Aristotle, in his extant works we do not find that he attributed the amber effect to Thales. In his work *On the Soul*, Aristotle mentioned that Thales ascribed a soul only to the magnet, as it can move the iron, but does not mention the amber effect explicitly:[10]

> Thales, too, to judge from what is recorded about him, seems to have held soul to be a motive force, since he said that the magnet has a soul in it because it moves the iron.

Archaeological excavations have shown that amber was known many centuries before Plato and even Tales.[11] It was utilized in jewelry and ornamentation. It is very probable that many people who worked with amber, traded it, or simply manipulated it, had observed its attractive properties, likely many centuries before Tales, although there is no written record supporting this conjecture.

In any event, it is known with certainty that at least from the time of Plato, in the 4th century B.C., the amber effect was known in ancient Greece. In antiquity, amber was probably rubbed with hair, tissue, or the skin of a person or animal. And it was observed that it attracted light bodies like a down-feather, chaff, or human hair.

2.3 Exploring the Attraction Exerted by Rubbed Bodies

In order to have a good standard of electrical neutrality for the next experiments, it is best to utilize at least two straws or two plastic rulers. One of them will

[8] [The56, p. 117].
[9] [The56, pp. 117-118] and [RR53].
[10] [Ari52c, A 2, Section 405, p. 634].
[11] [Gui05, p. 59].

never be rubbed. This will be our neutral straw. The other straw is the one which will be rubbed one or more times during the experiments. Even when this second straw has apparently lost its electrical charge between two experiments, it should not be utilized as a neutral straw due to the fact that some residual electric charge may remain on it. Sometimes just handling a piece of straw or removing dust which has accumulated on its surface can charge the straw. For this reason the criterion for a straw to be considered neutral must include the fact that it does not attract nearby light objects.

Experiment 2.2

We now repeat Experiment 2.1 rubbing the plastic straw in other materials such as a sheet of paper, the skin, a cloth, or a plastic bag. By moving the rubbed straw near small pieces of paper or chaff, we can see that they are attracted to the straw as in Experiment 2.1, when it was rubbed on hair, although not always with the same intensity. A plastic straw becomes highly electrified when rubbed on hair, with paper, or with a cotton tissue. Not always does it become so highly electrified when rubbed with a plastic bag.

2.4 Which Bodies are Attracted by the Rubbed Plastic?

Experiment 2.3

In this Section we want to answer the following question: Which bodies are attracted by the rubbed plastic? Questions such as these, or "is there repulsion?", "is there action and reaction?", "how many types of charge are observed in nature?", and the like are, of course, based on present knowledge. Normally the earlier investigators did not ask these questions or, at least, not in this form. These questions are more the result of their work than its motivation. In any event, we pose these questions in this book in order to draw attention to the main properties of electrical actions.

We now analyze which substances are attracted by a rubbed plastic or by a rubbed piece of amber. The rubbing can be done with paper, hair, or a cotton cloth. To do this, we place several groups of light substances in separate parts of a table. The substances can have been divided into small pieces, in short threads, or pulverized. Examples include chaff, thin cotton threads, small pieces of plastic bag, small pieces of aluminum foil (like the foil used in the kitchen or cigarette packages), chalk powder, flour, iron fillings, steel wool, small Styrofoam balls, down feathers, hair, sawdust, sugar, salt, small pieces of cork, etc.

We would like to emphasize one important point before carrying out these experiments. The rubbed object should not touch the substances on the table; it should only be brought near them. If the rubbed body touches these substances, they can stick to it due to moisture or to other sticky materials on the surface of the rubbed object or upon the surface of these substances, and not due to an electrostatic attraction between them.

When we move a neutral plastic near these substances nothing happens. After rubbing the plastic (or the amber) and moving it near these substances, without touching them, what is normally observed is that almost all of them are attracted to the rubbed plastic. That is, they move toward the rubbed plastic, jumping to it. Only the small pieces of plastic are not attracted, or are attracted very little in comparison to the attraction observed with the other substances.

Experiment 2.4

A similar experiment can be made with wires or threads of several substances: silk, cotton, polyester,[12] nylon (synthetic polyamide), hair, and copper. Low cost spools of silk, cotton, polyester, and polyamide threads can be found in sewing stores. The silk thread, in particular, will be employed in many experiments so it is a good idea to buy a spool bobbin. As for copper, it is possible to unwrap stranded wires sold at an electric shop. A stranded wire contains several thin copper wires side by side. In this experiment we utilize only a single one of these thin copper wires, cut into small pieces we will also refer to as threads.

We cut several pieces of these threads in the same length, for instance, 1 or 2 cm long. At one point on the table we place pieces of silk, at another, pieces of polyester, and so on. We move a neutral plastic near these substances and nothing happens. We rub another straw and bring it close to each one of these thread groups, without touching them. We observe that the cotton and copper threads are strongly attracted. The other substances are not attracted or are attracted much less than the cotton or copper threads.

In this case we have threads of the same length but of different weights, due to the different densities and thicknesses of these wires. However, it is easy to see that the cotton and copper are the heavier threads due to their higher density and sometimes also due to their greater thickness. Despite this fact, they are attracted more by the rubbed plastic than the lighter substances.

For substances which have approximately the same weight, we observe from these experiments that the majority of them are attracted by the rubbed plastic, although some of them are much more attracted than others. Only a very few of them do not seem to be attracted, or show a very weak attraction.

Experiment 2.5

An interesting experiment to show the attraction exerted by rubbed plastic upon metals utilizes empty aluminum beer or soft drink cans. One of these cans is laid upon a smooth surface. A plastic straw is rubbed and brought near this can, with the length of the straw held parallel to the can at the same height as the can's axis of symmetry. When the straw is very close to the can, without touching it, the can begins to move toward the straw (Figure 2.5). We can even make the can move forward and backward by changing the position of the rubbed straw, in front of or behind the can.

---

[12]See Appendix A.

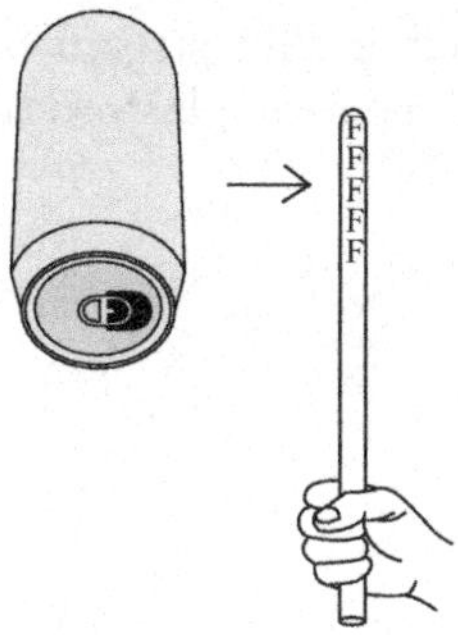

Figure 2.5: A rubbed straw attracts a metal can.

Experiment 2.6

We move a natural magnet, like a refrigerator magnet or a loudspeaker magnet, near the substances on the table described in Experiments 2.3 and 2.4. We observe that the magnet attracts only the iron fillings and the steel wool. The magnet does not affect the other substances—not even the pieces of copper wire or aluminum foil.

This is one of the main distinctions between electric and magnetic forces. Rubbed amber and rubbed plastic attract almost all light substances. A permanent magnet, on the other hand, attracts only a few substances: in general only those which contain iron.

The words magnet, magnetism, magnetic, etc. derive their name from a region called Magnesia, where the ancient Greeks found the naturally occurring magnetic mineral magnetite, an iron oxide which had the property of attracting small pieces of iron.

Experiment 2.7

We try to stick a magnet to an aluminum cooking pot, without success. We also do not succeed in attracting with the magnet the aluminum can of Experiment 2.5. This shows again the distinction between the electric and magnetic forces. It also confirms that not all metals are attracted by a magnet, but only a few types of metals, normally those which contain iron in their composition.

2.5 Is it Possible to Attract Liquids?

Experiment 2.8

In Experiment 2.1 we worked with solid substances. We now consider the effect of rubbed plastic upon liquids. Once more it is best to move the straw near the liquid. The straw can be neutral or it can have been rubbed previously. We should always prevent the straw from touching the liquid.

We turn on the tap so that a thin stream of water is running smoothly (Figure 2.6 (a)). We bring a neutral plastic straw close to the stream and nothing happens.

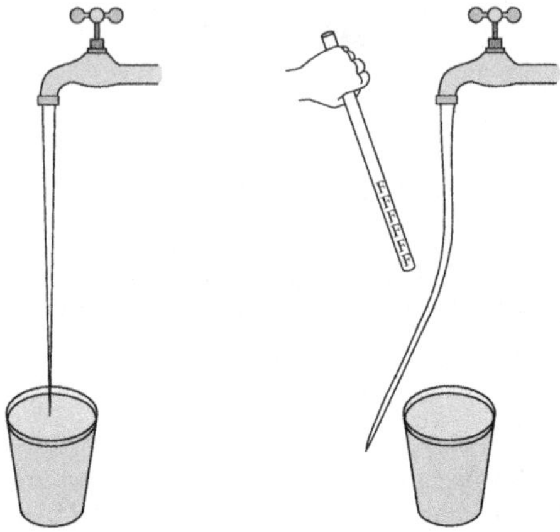

(a) (b)

Figure 2.6: A rubbed straw attracting a stream of water.

We rub another straw and bring it close to the stream. This time the stream bends toward the rubbed straw (Figure 2.6 (b)). This is more easily seen by moving the rubbed straw near the upper part of the stream, where the water has a lower velocity. Sometimes the attraction is so intense that the stream touches the straw. The experiment also works with dripping water. Once more the effect is more visible with the rubbed straw close to the slower droplets.

Experiment 2.9

Something analogous happens when we move a rubbed plastic near a stream of milk, detergent, alcohol, kerosene, shampoo, or kitchen vegetable oil. That is, all of these streams are attracted by the rubbed plastic, but do not feel any attraction by a straw which had not been previously rubbed. In the case of oil the effect—namely, the bending of the stream—is not so strong as in the case of the other liquids.

An experiment analogous to these seems to have been done for the first time by Jean Théophile Desaguliers (1683-1744) in 1741.[13] At the end of this paper Desaguliers said the following:

> Having properly suspended (that is, suspended by some electric body, here cat-gut)[14] a copper fountain with the spout downwards,

[13][Desb, pp. 666-667] and [Pri66, p. 85].

[14]See Appendix A.

I opened the cock, and let the water spout into a vessel underneath: Then, having excited [by friction] a great [glass] tube to electricity, I held it over the copper fountain, whilst an assistant held the thread of trial (that is, a thread hanging from a stick) near several parts of the jet, which attracted it sensibly: Then I applied the rubbed tube near to the falling jet, which attracted it strongly, so as to bend it into a curve, and sometimes cause it to fall out of the vessel below.

Students enjoy this funny and interesting experiment. It will be discussed in greater detail in Section 7.11.

Experiment 2.10

A similar experiment can be made with small puddles of water placed on a dry surface. When we bring a neutral straw near the water droplet, nothing happens. On the other hand, when we bring a rubbed plastic near the water, we observe that the water surface deforms, with the sections which are closest to the straw tending to approach the plastic (Figure 2.7). Sometimes the water droplet even flows over the dry surface, moving as a whole toward the rubbed plastic. The same happens with the other liquids mentioned previously, to a greater or lesser extent.

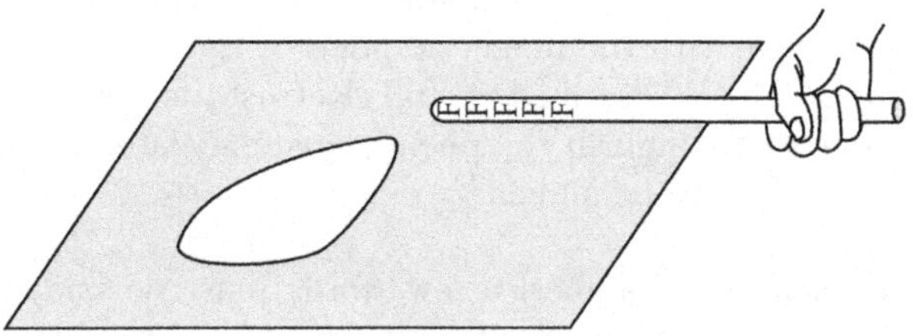

Figure 2.7: A rubbed straw attracting a water droplet.

2.6 Gilbert and Some of His Electrical Experiments

One of the scientists responsible for modern research on magnetism and electricity was William Gilbert (1544-1603) (Figure 2.8), an English medical doctor.[15]

In 1600 he published a very important book in the history of science, *On the Loadstone and Magnetic Bodies and on the Great Magnet the Earth.*[16] In this work he described many important discoveries related to magnetism. At that time the orientation of the magnetic needle was explained by an alignment of the magnetic poles of the needle with the poles of the celestial sphere. Gilbert,

[15] [Kel81].
[16] [Gil78].

Figure 2.8: William Gilbert (1544-1603).

on the other hand, presented the idea that the Earth is a huge magnet and thus has magnetic properties. He then explained the orientation of the magnetic needle by its alignment with the magnetic poles of the Earth.[17] In the second chapter of his book, Gilbert described several electrostatic experiments that were performed in order to distinguish the phenomena associated with the magnets from those associated with the amber:[18]

> Of this substance [the amber] a few words must be said, to show the nature of the attachments of bodies to it, and to point out the vast differences between this and the magnetic actions; for men still continue in ignorance, and deem that inclination of bodies to amber to be an attraction, and comparable to the magnetic coition.

He called *electrics* the bodies which had the same property as amber, our emphasis in italics:[19]

> The Greeks call this substance $\eta\lambda\epsilon\kappa\tau\rho o\nu$ [electron or amber], because, when heated by rubbing, it attracts to itself chaff; [...] These several bodies (*electrics*) not only draw to themselves straws and chaff, but all metals, wood, leaves, stones, earths, even water and oil; in short, whatever things appeal to our senses or are solid: yet we are told [by several ancient authors] that it attracts nothing but chaff and twigs.

[17][Kel81].
[18][Gil78, p. 27].
[19][Gil78, p. 27].

Or else:[20]

> And likeness is not the cause of amber's attracting, for all things that we see on the globe, whether similar or dissimilar, are attracted by amber and such like; hence no strong analogy is to be drawn either from likeness or from identity of substance.

Or also:[21] "A loadstone attracts only magnetic bodies; electrics attract everything."

Gilbert seems to have been the first to observe a liquid being attracted by a rubbed amber by performing an experiment analogous to that represented in Figure 2.7:[22]

> It [rubbed amber] plainly attracts the body itself in the case of a spherical drop of water standing on a dry surface; for a piece of amber held at suitable distance pulls toward itself the nearest particles and draws them up into a cone; were they drawn by the air the whole drop would come toward the amber.

The only exceptions to the attraction of rubbed amber mentioned by Gilbert were flaming or extremely rarefied objects,[23] "[...] for all bodies are drawn to all electrics, save bodies aflame or too rarefied, as the air which is the universal effluvium of the globe." He proved that rubbed amber does not attract air in the following way:[24]

> And that amber does not attract the air is thus proved: take a very slender wax candle giving a very small clear flame; bring a broad flat piece of amber or jet,[25] carefully prepared and rubbed thoroughly, within a couple of fingers' distance from it; now an amber that will attract bodies from a considerable radius will cause no motion in the flame, though such motion would be inevitable if the air were moving, for the flame would follow the current of air.

Later on he writes:[26]

> Electrics attract all things save flame and objects aflame, and thinnest air. And as they do not draw to themselves flame, so they have no effect on a versorium[27] if it have very near it on any side the flame of a lamp or of any burning substance; for it is plain that the effluvia are consumed by flame and igneous heat. Therefore electrics do not

[20] [Gil78, p. 28].
[21] [Gil78, p. 30].
[22] [Gil78, p. 31].
[23] [Gil78, p. 29].
[24] [Gil78, p. 31].
[25] See Appendix A.
[26] [Gil78, pp. 33 and 34].
[27] See Chapter 3.

> attract either flame or bodies near flame; for such effluvia have the virtue and analogy of rarefied humour, and they will produce their effect, bringing about unition and continuity, not through the external action of humours, or through heat, or through attenuation of heated bodies, but through the attenuation of the humid substance into its own specific effluvia. Yet they draw to themselves the smoke from an extinguished candle; and the lighter the smoke becomes as it ascends, the less strongly is it attracted, for substances that are too rare do not suffer attraction.

From what has been seen before, not all substances are affected by rubbed amber (or, at least, not all substances are attracted with the same strength). Even some substances having the same weight and shape are clearly more strongly attracted by a rubbed plastic than others. For example, equal threads of cotton or copper feel a stronger attraction than threads of silk or synthetic polyamide.

2.7 What Rubbed Substances Attract Light Bodies?

Experiment 2.11

We will now try to attract the small pieces of paper placed upon a table. We bring several rubbed objects near the pieces of paper. We will rub these objects in hair, in a sheet of paper, or in a cotton tissue. It is important to have homogeneous objects, that is, objects made of a single material, in order to avoid contradictory results. We should not, for instance, rub a plastic pen with metal parts. In this case it is best to rub a plastic straw and a metal spoon separately.

We list here some of these substances: plastic, amber, glass, wood, metal, acrylic, a natural magnet, thin cardboard, rubber, etc.

When the precautions mentioned previously have been taken, what is normally observed is that, after rubbing, amber, acrylic, and the plastic objects attract the small pieces of paper, as in Experiment 2.1 (Figure 2.3).

All the other substances do not normally attract the pieces of paper, no matter how long or how hard are they rubbed. This is represented in Figure 2.9 for a wood skewer.

In the case of glass there are exceptions, as there are several varieties of glass with varying compositions and made by different fabrication processes. But in general, after being rubbed, the most common glasses do not attract pieces of paper. The same can be said of rubber, as there are several varieties of rubber. The usual types of rubber found at home do not attract the pieces of paper.

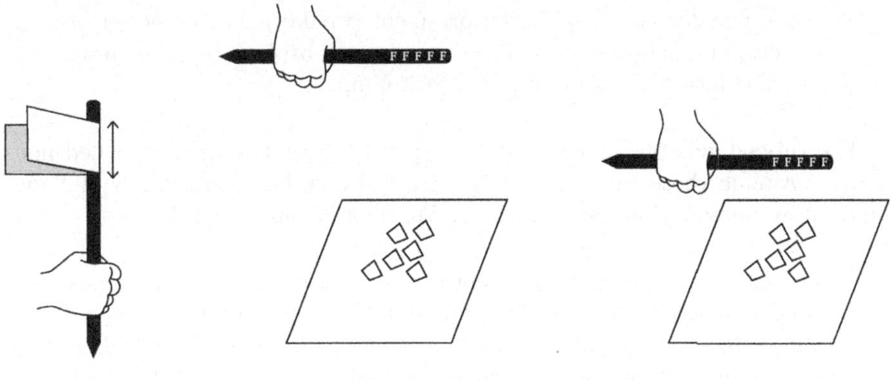

(a) (b) (c)

Figure 2.9: (a) A wood skewer being rubbed by a cotton tissue or in hair. (b) The rubbed skewer far away from pieces of paper. (c) We observe that it does not attract small pieces of paper when moved near them.

2.8 Gilbert's Nomenclature: Electric and Non-Electric Bodies

Until Gilbert's time only a few substances were known to attract small objects after being rubbed. They included amber, jet,[28] and diamond. It was in medieval times that it became known that jet, a hard compacted form of coal, also attracts like amber.[29] The natural magnet attracted iron and its composites. But it did not attract straws or chaff after being rubbed. The other substances also did not attract light objects after being rubbed. One of Gilbert's main contributions to the science of electricity was the discovery of many new substances which behaved like amber after being rubbed:[30]

> The ancients as well as moderns tell (and their report is confined by experience) that amber attracts straw and chaff. The same is done by jet, a stone taken out of the earth in Britain, Germany, and many other regions: it is a hard concretion of black bitumen,—a sort of transformation of bitumen to stone. [...] For not only do amber and (gagates or) jet, as they suppose, attract light corpuscles (substances): the same is done by diamond, sapphire, carbuncle, iris stone, opal, amethyst, vincentina, English gem (Bristol stone, *bristola*), beryl, rock crystal. Like powers of attracting are possessed by glass, especially clear, brilliant glass; by artificial gems made of (paste) glass or rock crystal, antimony glass, many fluor-spars, and belemnites. Sulphur also attracts, and likewise mastich, and

[28]See Appendix A.
[29][RR57, p. 546].
[30][Gil78, p. 27, Mottelay's words between square brackets].

> sealing-wax [or lac], hard resin, orpiment (weakly). Feeble power of attraction is also possessed in favoring dry atmosphere by sal gemma [native chloride of sodium], mica, rock alum.

The rubbed substances which did not attract light bodies were called non-electric. Among these substances Gilbert listed metals, several kinds of wood, the natural magnet, different gems, etc. We quote from his book:[31]

> In open air, heated objects cannot attract, not even metals or stones brought to a very high temperature by fire. For an iron rod at white heat, a flame, a candle, a flaming torch, or a red-hot coal when brought near straws or to a revolving pointer (*versorium*) does not attract; and yet plainly all these cause the air to come to them in a current, for they consume air as a lamp consumes oil.

The following list is very important:[32]

> But very many electric bodies (as precious stones, etc.) do not attract at all unless they are first rubbed; while sundry other bodies, and among them some gems, have no power of attraction, and cannot be made to attract, even by friction; such bodies are emerald, agate, carnelian, pearls, jasper, chalcedony, alabaster, porphyry, coral, the marbles, lapis lydius (touchstone, basanite), flint, bloodstone, emery or corundum, bone, ivory; the hardest woods, as ebony; some other woods, as cedar, juniper, cypress; metals, as silver, gold, copper, iron. The loadstone, though it is susceptible of a very high polish, has not the electric attraction.

Likewise:[33]

> For this reason it is that neither metals, marbles, flints, woods, grasses, flesh, nor various other substances can attract or solicit a body, whether magnetically or electrically (for it pleases us to call electric force that force which has its origin in humours). But bodies consisting mostly of humour and not firmly compacted by nature wherefore they do not stand friction, but either fall to pieces or grow soft, or are sticky, as pitch, soft rosin, camphor, galbanum, ammoniacum, storax, asa, gum benjamin, asphaltum (especially in a warm atmosphere), do not attract corpuscles. For without friction few bodies give their true natural electric *emanation* and effluvium. Turpentine resin in the liquid state does not attract, because it cannot be rubbed; but when it hardens to a mastic it does attract.

[31] [Gil78, p. 28].
[32] [Gil78, p. 29].
[33] [Gil78, p. 30].

Several words used to this day have their origin in the word amber (or *electron* in Greek): electric, electron, electricity, electret, electronic, electrician, electromagnet, electrode, etc. Originally the word *electricity* meant the property or the power to attract light bodies, as was the case for rubbed amber.[34] This word occurred for the first time in a printed work by Sir Thomas Browne (1605-1682) in 1646. In 1820 Ørsted introduced the terms *electromagnetism* and *electromagnetic*, while in 1822 Ampère introduced the terms *electrostatics* and *electrodynamics*.[35]

Gilbert called all bodies which attracted light substances after being rubbed *electric*, though this nomenclature is no longer in use. The reasons for this change of nomenclature are given in Chapter 6, Chapter 8, and Appendix B. In order to aid in understanding several historical quotations that will appear in this book, it is important to be aware that these materials are now classified as *insulators* and *conductors*. Insulators are also called *nonconductors* or *dielectrics*. The substances Gilbert classified as electric are now called insulators. And the substances which were previously classified as non-electric are now called conductors.

[34][RR57, p. 558], [Hea67], and [Hei99, p. 169].

[35][Amp22, p. 60], [Ørs98a, p. 421], [Ørs98b, p. 426], [Blo82, p. 78], [GG90, p. 920], [GG91, p. 116], and [Cha09, pp. 24-26].

Chapter 3

The Versorium

3.1 Fracastoro's Perpendiculo and Gilbert's Versorium

We will now discuss the oldest electrical instrument. It was created by Girolamo Fracastoro (1478-1553) (Figure 3.1). Some give his name as Fracastoro, others as Fracastorio.[1] He was a poet, physician, and philosopher in Verona.[2] Fracastoro is better known for his works on medicine, especially epidemiology. He gave the name *syphilis* to a known venereal disease.

Figure 3.1: Girolamo Fracastoro (1478-1553).

[1] [Ben98, p. 241].
[2] [Zan81].

His instrument was first presented in a book published in 1546.[3] He utilized it to show that rubbed amber attracts not only straws and chaff, but also another piece of amber, and even a metal like silver. He also discovered that diamond has the property of attracting light substances after being rubbed, as is the case with amber. Fracastoro describes his new instrument with the following words:[4]

> In fact we, in the presence of several of our medical doctors, have made many experiments with a *perpendiculo* which is well adapted as in a marine compass, and have observed how a magnet attracts another magnet, [magnetized] iron [attracts] iron, due to the fact that a magnet attracts iron and the iron [attracts] the magnet; moreover, [rubbed] amber snatches up little pieces of amber... and, likewise, [rubbed] amber attracts to itself not only straws and chaff, but also silver.

When he writes *perpendiculo*, Fracastoro might be referring to a plumb line,[5] i.e., a small object suspended by a vertical thread from a support, like a pendulum. The thread would be free to move in any direction around the point to which it is attached. The word "perpendiculo" is connected with "perpendicular," which means a straight line at right angles to the horizon. A plumb line is utilized to indicate a vertical direction. For this reason it is natural to suppose that Fracastoro's perpendiculo was analogous to a plumb line.

From description above we infer that Fracastoro attached a small piece of amber or silver to the end of the thread. When he brought a rubbed amber near the perpendiculo, he would have observed that it departed from the vertical direction, moving closer to the rubbed amber (Figure 3.2). The advantage of the perpendiculo in comparison with straws or chaff; is that the tension of the thread conterbalances the weight of the suspended body. It is then easy to see its motion in a horizontal direction even for a small attractive force. On the other hand, if the small piece of amber or silver was on a table, it would have been more difficult to observe or detect any motion due to its weight. That is, it would be difficult to see its vertical motion toward a rubbed amber placed near it.

Gilbert knew Fracastoro's book and quoted it several times in his book:[6]

---

[3] [Gli33] and [Hei99, p. 175].

[4] "Nos enim praesentibus multis è nostris medicis experientiam multorum secimus, perpendiculo bene & concinne aptato, quale est in nauigatoria pyxide, ac manifeste vidimus magnete trahere magnete, ferrum ferrũ, tum magnetem trahere ferrum, ferrum magnetem porro electrum parua electri frustula rapere, argentum attrahere argentum, &, quod valde inirati fui mus, magnetem vidimus argentum trahere: item Electrum non solum furculos & paleas mouere ad se, sed & argentum," [Fra55, p. 85 verso]. In Italian: "Noi infatti alla presenza di molti dei nostri medici facemmo esperienza di molte cose con un perpendiculo bene e convenientemente adattato come è nella bussola da navigare e vedemmo manifestamente che il magnete attrae il magnete, il ferro il ferro, poi che il magnete attrae il ferro e il ferro il magnete; e ancora, l'ambra rapisce pessettini d'ambra... e parimenti l'ambra non avvicina solamente a sè i fuscelli e le pagliuzze, ma anche l'argento," [Gli33].

[5] [Sas02].

[6] [Gil78, pp. 28-29].

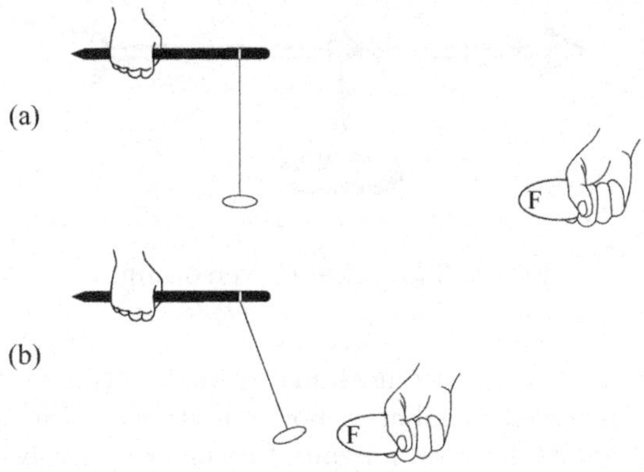

Figure 3.2: Possible representation of Fracastoro's perpendiculo and the experiment he may have been done with it. (a) The hand holds a large piece of amber. The rubbed section of this amber is represented by the letter F. At the lower end of the perpendiculo there is another small piece of amber or silver which has not been rubbed. When the large piece of amber is far away from the perpendiculo, the string remains at rest vertically. (b) The perpendiculo is attracted when the rubbed amber F is brought close to the small piece of amber or silver.

> Fracastorio thinks that all bodies that mutually attract are alike, or of the same species, and that, either in their action or in their proper *subjectum*: "Now the proper *subjectum*," says he [Fracastoro], "is that from which is emitted that emanational something which attracts, and, in mixed substances, this is not perceptible on account of deformation, whereby they are one thing *actu*, another *potentia*. Hence, perhaps, hairs and twigs are drawn to amber and diamond not because they are hairs, but because there is imprisoned within them either air or some other principle that is first attracted and that has reference and analogy to that which of itself attracts; and herein amber and diamond are as one, in virtue of a principle common to both." So much for Fracastorio.

Gilbert probably began to investigate the attractive properties of other precious stones after studying this book by Fracastoro. Gilbert also describes an instrument which he called a *versorium*,[7] though he did not mention that a similar instrument, the *perpendiculo*, had been invented by Fracastoro. Gilbert's original image of this instrument is shown in Figure 3.3.

The name *versorium* comes from a Latin word having the meaning *to turn* or *to revolve*. The versorium is an instrument which normally consists of two parts: a vertical member, which acts as the support, and a supported horizontal member which is capable of freely turning around the vertical axis defined by the

[7] [Gil78, pp. 27-28].

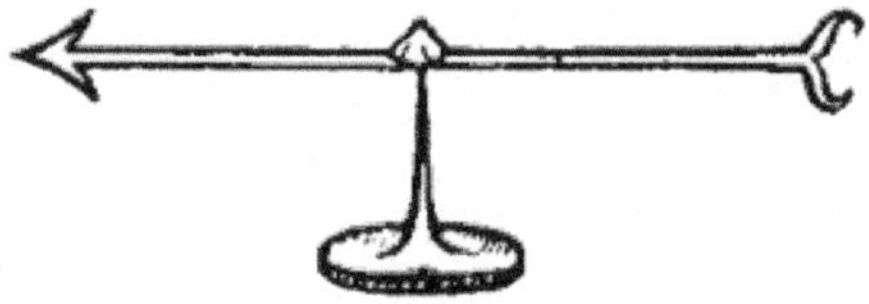

Figure 3.3: Gilbert's versorium.

support. In this respect, it is very much like a common compass in construction except that the horizontal member is not magnetized as it is in a compass. Conceptually, the ability for the horizontal member to freely rotate means it is very sensitive to extremely small external torques and hence may be used to detect them in the same way a compass detects the magnetic torque of the Earth.

When at rest it will point in an arbitrary horizontal direction (it can point along the East-West direction, for instance, or toward a tree).

3.2 Making a Versorium

There are three different ways to build a versorium.

3.2.1 Versorium of the First Kind

The versorium of the first kind is like Gilbert's versorium. It can be built by attaching a pin, toothpick, or nail with its pointed tip upwards from a rigid base. The base should be heavy or attached to a table to prevent the entire instrument from falling. The vertical support can be a cork with a pin, a toothpick standing vertically in modeling clay, or a thin board with a nail. The only requirement is that the support should remain fixed relative to the ground, while the horizontal member is free to turn in a horizontal plane above the vertical axis formed by the support. The mobile horizontal piece is supported at its center by the tip of the pin.

It is important to note that in order to prevent the mobile part from falling, it is essential that its center of gravity should be located below the point of contact between it and the pointed tip of the vertical support. A detailed discussion of the center of gravity (CG) and the experimental procedures to find it can be found in the book *Archimedes, the Center of Gravity, and the First Law of Mechanics.*[8]

There are several ways to set the CG of a mobile part below its point of contact with the pin. For example, the mobile part can be in the shape of an upside down letter V, or it can have its center (which will be in contact with the pin) bent upward in such a way that when it is set on the pin, the tip of the pin is located above the plane of the flat mobile part. A simple mobile

[8] [Ass08a] and [Ass08b].

part can be made with a brass or steel paper fastener. In this case it is best to bend the center of the circular base of the paper fastener a little. This bent portion will be supported on the pin. To bend the paper fastener we utilize a nail and a hammer, but carefully, without making a hole in the top of the fastener, only bending it a little to create a small indentation. The mobile part will be supported by this bent section placed on the tip of the pin in such a way that it will not slip off the pin. After the legs of the paper fastener have been bent downward so that it makes an upside down letter V, the fastener can be set on the pin.

The mobile part may also be made using an aluminum strip (which can be obtained cutting a soft drink can), a dry straw, wood, thin cardboard, or piece of plastic (a hard plastic strip). The important thing is to shape the mobile part into an inverted letter V. The hard plastic can also be folded so the two legs point downward. When the mobile part is placed on the pin, it is important to verify that it has complete freedom to rotate clockwise and counterclockwise in a horizontal plane, without slipping or sticking due to friction with the pin. It is then ready for the experiments.

The versorium of the first kind is depicted in Figure 3.4. In (a) we have the base for the versorium (in this case a pin set in a cork). The mobile part is shown in Figure 3.4 (b). In this case it is a steel paper fastener seen from above and the side, with the center of its head a little bent and its legs inclined downward. The complete mounted versorium, with the center of the fastener set on the tip of the pin, is shown in Figure 3.4 (c).

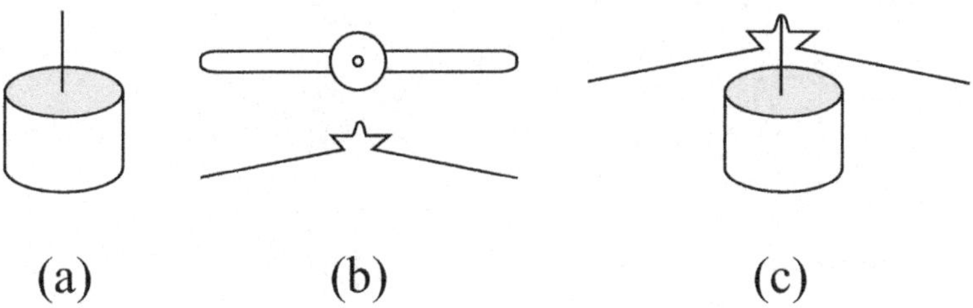

Figure 3.4: Versorium of the first kind. (a) Versorium base. (b) Steel paper fastener seen from above and the side. (c) The mounted versorium.

3.2.2 Versorium of the Second Kind

The second way to make a versorium is by attaching a pin to the horizontal mobile part of the versorium. We will call this mobile part the "hat," which can be a plastic or metal strip. The pin is securely attached through the center of the hat, with the tip of the pin pointing downward. The pin rotates together with the hat. This system is then supported on a small horizontal flat surface which is fixed relative to the ground, like the head of a nail stuck in a board or cork. In Figure 3.5 we see a representation of this kind of versorium. (a) Its base, in this case a nail stuck in a board. (b) The mobile part of the versorium, in this case a strip of plastic or metal with a pin attached to its center, with its tip downward. (c) The complete versorium, with the tip of the pin set on the horizontal head of the nail stuck in a board.

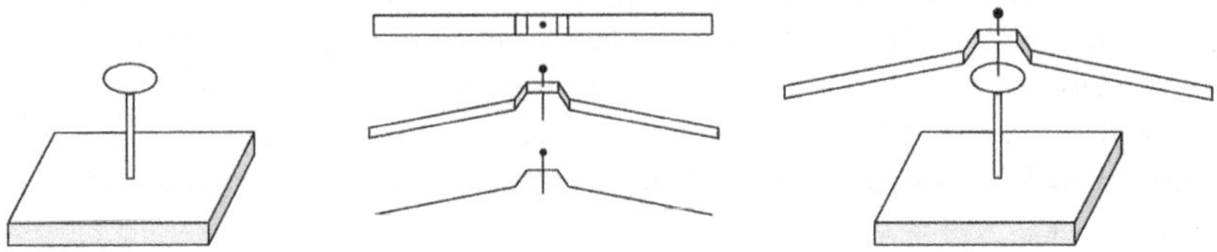

(a) (b) (c)

Figure 3.5: Versorium of the second kind, with the pin set on the mobile part of the versorium. (a) Fixed base of the versorium. (b) Hat of the versorium (plastic or metal strip) with the pin attached to it. (c) Mounted versorium.

In order to prevent the versorium from slipping, it is crucial that the center of gravity of the hat and pin be lower than the tip of the pin. The center of gravity of the pin only is located at a point A between the head H and the tip T of the pin (Figure 3.6 (a)). Normally this point A will be closer to H than to T, although we show it here close to the center of the pin. The center of gravity of the hat only is at a point B along its vertical axis of symmetry, between its top and bottom parts (Figure 3.6 (b)).

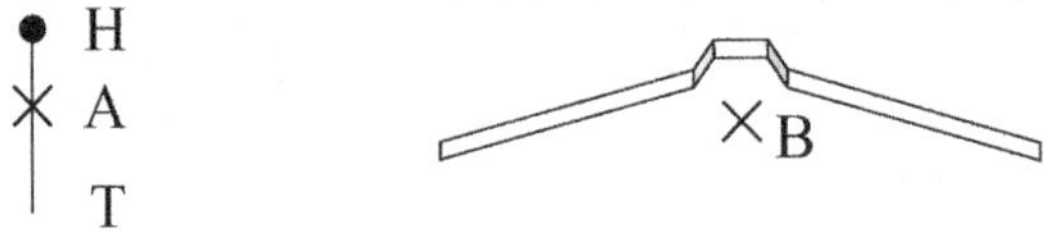

(a) (b)

Figure 3.6: (a) The center of gravity of the pin is A. (b) The point B is the center of gravity of the hat.

The center of gravity of the entire mobile part (hat and pin) of this kind of versorium is located at a point C between A, the center of gravity of the pin, and B, the center of gravity of the hat. Three possibilities exist as shown in Figure 3.7. (a) If the pin has the same weight as the hat, then C will be at the midpoint between A and B. (b) If the pin is heavier than the hat, C will be closer to A. (c) If the pin is lighter than the hat, C will be closer to B.

If C is higher than the tip T of the pin, the versorium will slip off the nail making it not possible to balance it above the nail. The reason is that the mobile part of the vesorium will be in unstable equilibrium in this configuration. In Figure 3.8 (a) we illustrate this situation of unstable equilibrium with the point C (from Figure 3.7) represented by the symbol $\times$. Here $\times$ is vertically

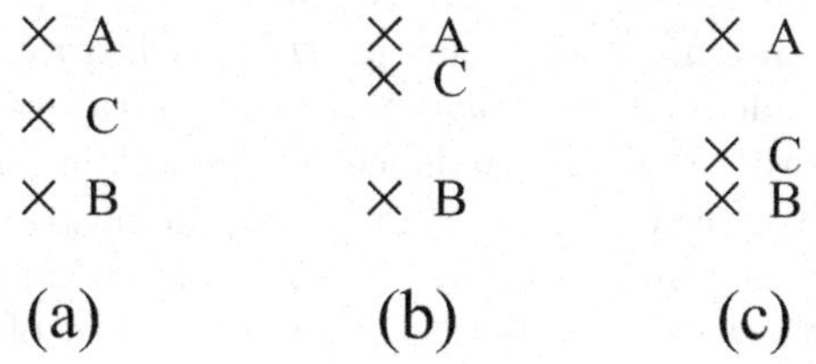

Figure 3.7: Location of the center of gravity C of the mobile part of the versorium. (a) Pin and hat with the same weight. (b) Pin heavier than the hat. (c) Pin lighter than the hat.

above T, at its highest position. Let us suppose that the mobile part deviates slightly from this unstable situation, that is, the mobile part dips a little in a clockwise or counterclockwise direction around the tip T of the pin, which lowers one of its legs while raising the other. In this case the center of gravity $\times$ of the mobile part will move below its initial position. The tendency of the center of gravity of any system is to approach the Earth's surface when this possibility exists. Therefore, the versorium will continue rotating in the clockwise or counterclockwise direction, resulting in the mobile part falling down.

The only way to balance the mobile part of the versorium above the nail is to have point C below the tip T of the pin. This is represented in Figure 3.8 (b), with the symbol $\times$ meaning the position of the center of gravity of the mobile part of the versorium (made up of the pin together with the hat). This is the configuration of stable equilibrium, with $\times$ in its lowest position, namely, vertically below the tip T of the pin. In this stable configuration, any motion of the versorium in the clockwise or counterclockwise direction around the tip T of the pin, will raise the center of gravity $\times$ in comparison with its height when it was vertically below the tip T. The system will then return to the configuration of stable equilibrium due to the gravitational restoring torque exerted upon it by the Earth.

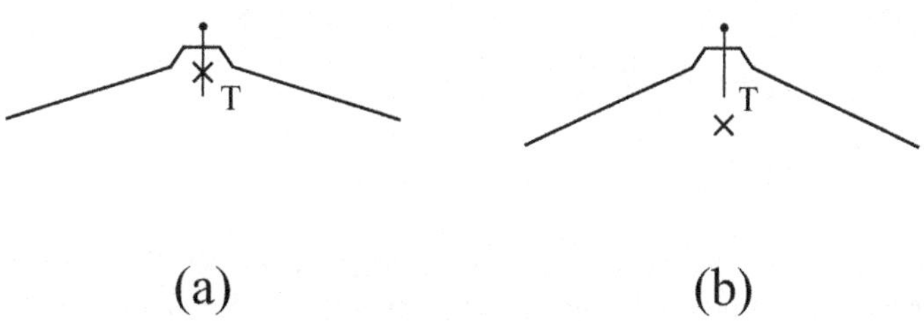

Figure 3.8: The symbol $\times$ represents the center of gravity C of the mobile portion of the versorium (composed by the pin and the hat). (a) Versorium of the second kind in unstable equilibrium, with $\times$ above the tip T of the pin. (b) Versorium in stable equilibrium, with $\times$ below the tip T.

Sometimes it is difficult to obtain this configuration of stable equilibrium with a light mobile part made of a plastic straw. To prevent this problem we can use one straw inside another straw, or a plastic strip made of a denser and heavier material, in order to counterbalance the pin weight. Another alternative is to cut the heavier top part of the pin (including the head) with pliers, keeping only the lower part (including the tip). You can also bend the legs of the hat of the versorium downward in order to lower its center of gravity or simply use longer hats. Another alternative is to replace the pin with a small nail passing through the center of the mobile part, or stuck to its center with glue or modeling clay. When the system is ready, it is important to test whether it is free to rotate in both directions in a horizontal plane around a vertical axis without slipping. If it slips sideways, you can balance it by lowering one of the legs, or increasing the length of one of the legs. We are then ready to begin the experiments.

3.2.3 Versorium of the Third Kind

The third way of making a versorium is perhaps the simplest one. For the mobile part we choose a strip made of metal, wood, or another appropriate material and attach a cotton or silk thread to its center. The strip should remain horizontal when at rest and tied by its center. We then fasten the upper end of the thread to a support which is fixed relative to the ground. The mobile part attached the lower end of the thread is then free to rotate horizontally in both directions around the vertical thread. Figure 3.9 illustrates this type of versorium with the mobile part supported at its center by a vertical thread attached to a pencil. Fracastoro's *perpendiculo* was probably a versorium of this kind.

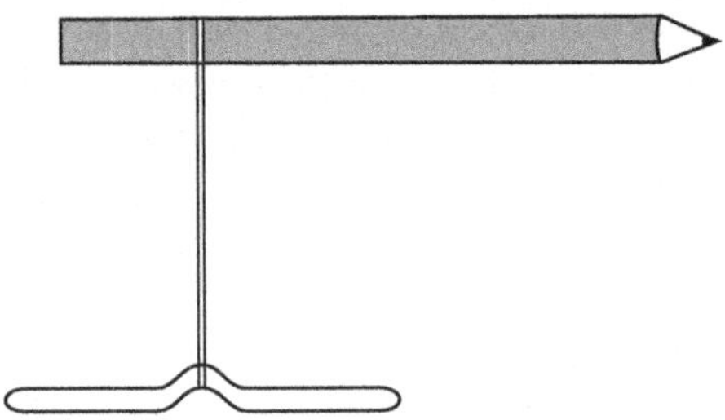

Figure 3.9: Versorium of the third kind.

The versorium of the third kind has a property which differentiates it from the other two kinds. The mobile part of the versoria of the first and second kinds can only incline or rotate around their centers which remain at rest relative to the ground. The versorium of the third type, on the other hand, cannot only rotate around a vertical axis, but like a pendulum it can also move as a whole when attracted by another body. This has an advantage in terms of the versatility of its motion. However, sometimes this complicates the analysis of the phenomena which we wish to describe or observe. In the following experiments we initially use only versoria of the first and second kinds.

Given a constant external torque, it is easier to rotate a lighter weight mobile part of a versorium than a heavier one. This means that a light versorium has a greater sensitivity than a heavy one.

Although Gilbert built only metal versoria, they can be made of different materials: metal, plastic, thin cardboard, dry straw, wood, etc. Initially we will work only with metal versoria, which we will call simply versoria. When the spinning needle is made of plastic, paper, or of another non-metal material, we will call the system a plastic versorium, a paper versorium, or the appropriate name. In this way we will be able to distinguish these versoria from the versorium utilized by Gilbert.

3.3 Experiments with the Versorium

Experiment 3.1

We bring a neutral plastic near a metal versorium, without bringing them into contact. Nothing happens (Figure 3.10).

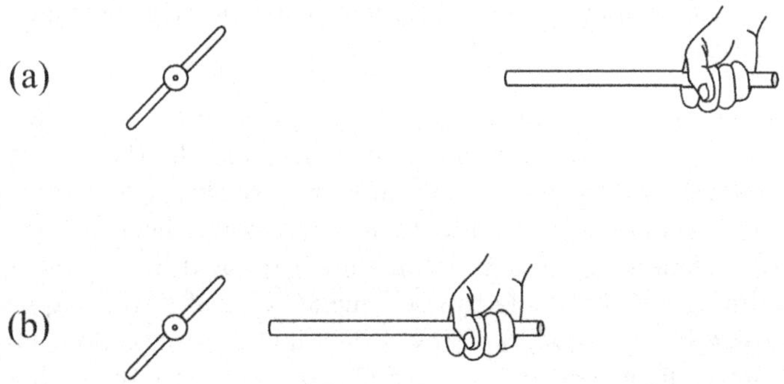

Figure 3.10: (a) A metal versorium points in an arbitrary direction when it is far away from a neutral piece of plastic. (b) The versorium remains at rest when the neutral plastic is brought near it.

We rub another piece of plastic and repeat the experiment. In this case we observe that the metal versorium is oriented by the rubbed plastic, pointing toward it (Figure 3.11). The same happens with a wood versorium and a paper versorium.

This experiment shows that the rubbed plastic affects nearby bodies, as we saw in Experiment 2.1. But there are two main differences between these two experiments. The first difference is that in Experiment 2.1 there was movement of the small pieces of paper. In the present experiment only the versorium changes direction, while its center remains at rest above the pin. The second difference is that the versorium moves more easily than the pieces of paper. That is, some rubbed objects cannot attract light bodies to themselves. However,

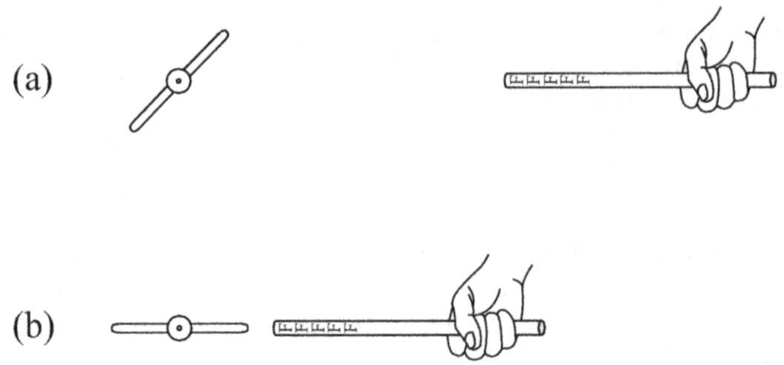

Figure 3.11: (a) A metal versorium points in an arbitrary direction when it is far away from a rubbed piece of plastic. (b) When the plastic is brought close to it, the versorium is oriented by it, pointing constantly toward the plastic.

these same rubbed objects can move metal versoria. The versorium is a better detector of weak electrification than pieces of paper or straw. Gilbert utilized this great sensitivity to discover many new electric objects, i.e., rubbed objects which can attract or orient other substances placed in their neighborhood.

Gilbert described the versorium as follows:[9]

> Now in order clearly to understand by experience how such attraction takes place, and what those substances may be that so attract other bodies (and in the case of many of these electrical substances, though the bodies influenced by them lean toward them, yet because of the feebleness of the attraction they are not drawn clean up to them, but are easily made to rise), make yourself a rotating-needle (electroscope—*versorium*) of any sort of metal, three or four fingers long, pretty light, and poised on a sharp point after the manner of a magnetic pointer. Bring near to one end of it a piece of amber or a gem, lightly rubbed, polished and shining: at once the instrument revolves.

The word *electroscope* in this quotation was introduced by Mottelay in his English translation of Gilbert's work. It does not appear in the original Latin text, in which only the word *versorium* is utilized.[10] The word *electroscope* does not appear as well in Thompson´s translation of Gilbert's book.[11] Mottelay utilized the word "electroscope" with the meaning that this instrument might indicate, by its orientation, which objects behaved like amber after being rubbed. *Electroscope* is the generic name of any device which is sensitive enough to detect a force or torque of electrical origin. In this book, on the other hand, we will reserve the name electroscope specifically to the instrument discussed in Section 6.1.

[9] [Gil78, pp. 27-28].
[10] [Gil00, pp. 48-49 and Glossary] and [Hea67].
[11] [Gil00, pp. vj and 48-49] and [Hea67].

We have now a second criterion in order to say that a plastic or another material is *electrically neutral*. The first criterion was presented in Experiment 2.1, namely, not to attract lightweight bodies. The second criterion is that of not producing orienting motion of a metal versorium when this material is placed close to one of the legs. In the following experiments it is important to keep one neutral plastic straw or ruler which neither attracts light bodies nor orients metal versoria. This straw or ruler should not be rubbed in any event, as it will be utilized as our neutral standard.

Fracastoro did not describe how he created the perpendiculo, which came before Gilbert's versorium. We can only speculate as to how he made it. One of the goals of his book was the study of magnetism. He also wanted to distinguish amber attraction from the attraction exerted by a natural magnet. Perhaps he did rub a piece of amber in order to perform an electrical experiment and casually noticed that it was capable of rotating a compass needle. As amber is not magnetic, whether it is rubbed or not, he should have concluded that this orientation of the compass was due to an electric attraction, analogous to the attraction of lightweight bodies by the rubbed amber. He could have decided then to make metal needles analogous to magnetic needles, but not magnetized. They would rotate toward a rubbed piece of amber, but not toward a magnet (assuming needles made of copper or silver, but not made of iron nor steel). He may thus have created the first artificial instrument for the study of electricity.

Experiment 3.2

We move a magnet near a versorium. We see that only versoria made of steel, iron, nickel, or other ferromagnetic materials rotate and direct themselves toward the magnet. Versoria made of other materials are not affected by the magnet. We also see that several metals, like copper and aluminum, are not affected by the magnet. The same happens with the majority of substances (paper, plastic, wood, etc.)

With this experiment we can distinguish magnetic interaction from electric interaction, as we did earlier with Experiments 2.6 and 2.7, but now with a greater precision.

3.4 Is it Possible to Map the Electric Force?

Is it possible to map the electric force exerted by a rubbed plastic body? Can we visualize in which direction a long rubbed straw will attract a piece of paper placed nearby? In this Section we answer this question.

In the following experiments we can use several versoria simultaneously. We can also use a single versorium which is placed alternately in several positions around the rubbed body for each experiment. In the next figures we show several versoria at the same time. It is best to utilize small versoria, such as the ones made of small paper fasteners placed on pins. These pins can be attached to several corks, or they can all be stuck into a Styrofoam board. Initially we work only with metal versoria.

Experiment 3.3

We move a neutral straw near the versoria, nothing happens. We rub the tip of another plastic straw. We place this rubbed tip at the same height as the plane formed by several versoria upon a table. We observe that they turn and point toward the rubbed tip of the straw (Figure 3.12). In this figure the central circle with the letter F indicates the rubbed tip of the straw. The influence of the rubbed straw extends about 10 cm. The more distant versoria are not apparently affected by the rubbed straw, unless it is placed close to them.

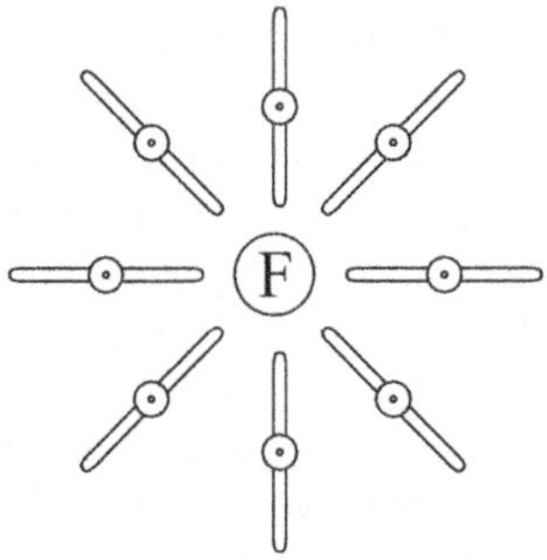

Figure 3.12: The nearby versoria are oriented toward the rubbed tip of the plastic.

The orientations indicated by the versoria represent the directions of the electric force exerted by the rubbed plastic. That is, if there are pieces of paper in the locations of the versoria, and if the attractive force of the rubbed straw is strong enough, the orientation of the versoria indicate the directions of motion which would be produced upon the pieces of paper due to the presence of the rubbed plastic. This means that they would be radially attracted toward the rubbed tip.

The versoria in this experiment function like iron fillings spread around a permanent magnet, indicating the directions of the magnetic forces exerted by the magnet upon other magnetic poles or upon small iron pieces.

Experiment 3.4

Analogous experiments can be made for different configurations. For instance, we can rub a plastic straw along its length, and then stand it vertically on a base, such as around a toothpick stuck in modeling clay. The nearby versoria will point toward the rubbed straw.

We can also support this rubbed straw horizontally by attaching it at the ends. The final configuration for the versoria in this case is like that of Figure 3.13. Most versoria will point toward the rubbed plastic, while the versoria that are closer to its ends will point toward these ends.

Experiment 3.5

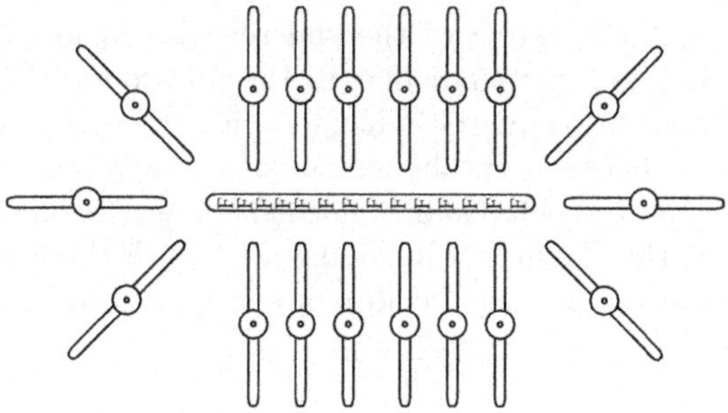

Figure 3.13: Orientation of versoria toward a horizontal straw which has been rubbed along its entire length.

We now repeat these experiments using two rubbed straws standing vertically. In this case the configuration of the versoria is shown in Figure 3.14. The circles with the letters F represent the rubbed portions of the straws which are at the same level as the versoria. This configuration indicates a vectorial addition of the torques exerted by each plastic straw on the versoria. Vectors add to produce the resultant by the parallelogram rule.

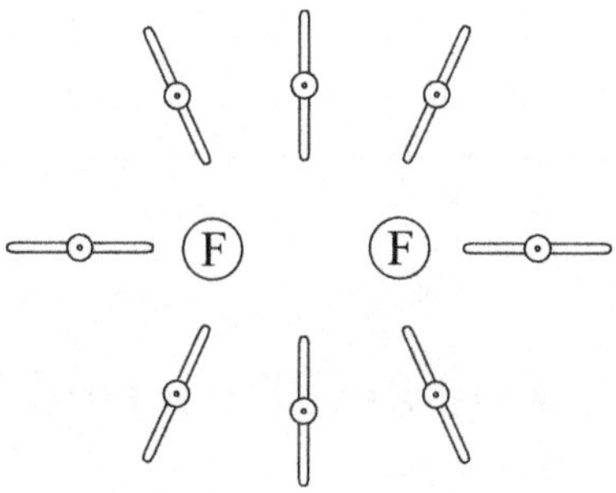

Figure 3.14: Orientation of the versoria due to two rubbed plastic straws.

It is of historical interest that Gilbert did not employ a versorium in order to map the electric force as we are doing here. But he did use magnetized compass needles in order to map the magnetic force of a magnet. In Figure 3.15 we show the results he obtained for cylindrical and spherical magnets.[12] The spherical magnet orients the compasses analogously to the orientation of compasses above the Earth, which point toward the North-South magnetic poles. In other words, the small spherical magnet also has two poles, which are the points upon the surface of the sphere close to which the compasses remain perpendicular to the surface of the sphere, pointing toward the center of the sphere. It is possible to draw the magnetic meridians over the surface of this sphere. They are circles connecting these two poles, with the centers of the

[12][Gil78, pp. 10 and 82].

circles coinciding with the center of the sphere. The magnetic equator is the great circle with its plane perpendicular to the line connecting the two poles. The center of the magnetic equator coincides with the sphere's center. Gilbert utilized this analogy between the behaviour of small compasses near a small magnetized sphere and the behaviour of normal compasses on the surface of the Earth to argue that the Earth is a huge magnet. With this model he was able to justify the orientation of normal compasses used in terrestrial navigation.

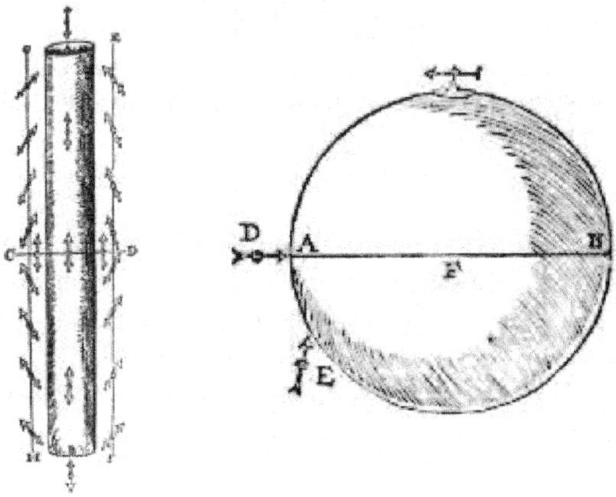

Figure 3.15: Mapping of magnetic force done by Gilbert using magnetized needles brought close to cylindrical and spherical magnets. The poles of this spherical magnet are located at A and B.

3.5 Is There Action and Reaction in Electrostatics?

Thus far we have seen that rubbed amber, or rubbed plastic, attracts and moves lightweight objects and causes versoria to turn. We now will analyze the opposite process.

Experiment 3.6

A neutral plastic straw is placed in contact with a wall and released from rest. It falls to the ground (Figure 3.16 (a)). We rub another straw along its entire length with a piece of paper or some hair. We then place it in contact with the wall and release it from rest. We see that it remains stuck at the wall despite the gravitational attraction of the Earth (Figure 3.16 (b)). The same effect is observed when it touches a glass window, a piece of metal furniture, or a school blackboard. It can even stick to the ceiling!

This experiment can also be used to indicate when a straw is well electrified. If it sticks on the wall after being rubbed, it has good electrification. If it slips or

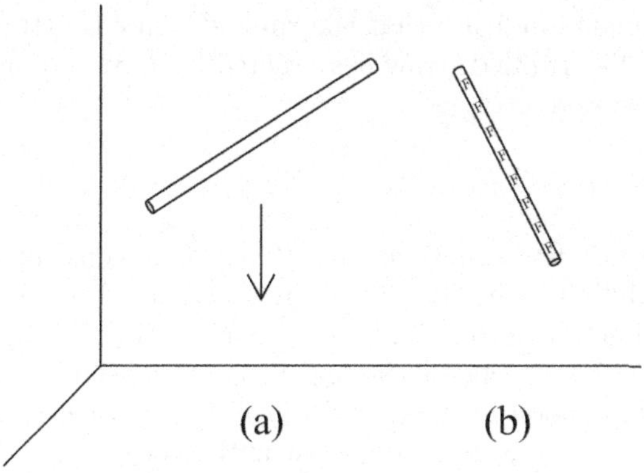

Figure 3.16: (a) A neutral plastic straw falls to the ground after release. (b) A plastic straw rubbed along its entire length remains stuck on a wall after being released from rest.

falls to the ground very soon after being rubbed and released, this indicates that it has weak electrification. Most experiments in this book will work if we use well electrified plastic straws. This wall test can be used to discover materials that can store a large amount of electricity, and also the best or most efficient ways to apply friction. For instance, how do we obtain better electrification: by rubbing a straw in hair, in a piece of paper, or in a plastic bag? This wall test is a very useful, practical, and simple method to check the electrification of a plastic straw.

Sometimes the rubbed straw does not stick on the wall even after a large amount of friction. This may happen with dense, heavy straws. In this case, the weight of the straw will be greater than the electric force, and also larger than the force of friction between the rubbed straw and the wall. As a result, the straw will fall to the ground. When this happens, it is best to change to another kind of straw that is thinner and lighter.

The same test can be performed with a piece of plastic bag, instead of a straw. A neutral piece of plastic bag will fall to the ground after touching a wall and being released from rest. On the other hand, a rubbed piece of plastic bag will stick to the wall for a long time.

Experiment 3.7

An analogous experiment can be performed with an inflated rubber balloon. When we touch it against a wall and release it from rest, it falls to the ground. We now rub the balloon against hair. We touch the rubbed section of the balloon to a wall and release it from rest. If it is well electrified, it will stick to the wall after release. For this experiment to succeed the balloon should be rubbed briskly over a large area.

These are very simple experiments, but the results are impressive. Sometimes the straw can remain stuck on the wall for several minutes or even an hour. The experiments indicate that the rubbed straw is attracted by the wall or by the ceiling. The rubbed straw is attracted by several different substances: wall, glass, metal, wood, etc.

Experiment 3.8

We now perform a few experiments with a plastic versorium, like the versorium of the second kind. Initially we work with a neutral versorium which is free to turn in both directions around a vertical axis passing through its center. We bring several objects near one of the legs of the versorium. These objects can be a finger, a metal spoon or wire, a wood barbecue skewer, a sheet of paper, or a piece of cloth. Each body is moved near the versorium separately from all the others. Nothing happens with the versorium. That is, its previous arbitrary orientation is not affected (Figure 3.17).

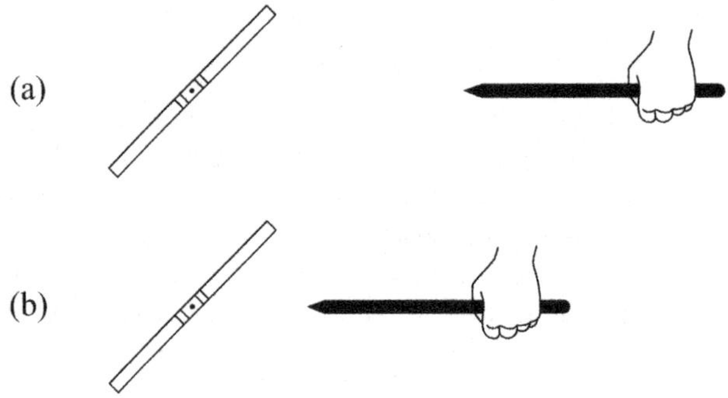

Figure 3.17: A neutral plastic versorium is not oriented when a finger, a piece of metal, wood, paper, or cloth is brought near one of its legs.

We now rub only one of the legs of the plastic versorium with a sheet of paper or a cloth. We repeat the procedure when we bring a finger near this rubbed leg. We can also move a piece of metal or wood near the rubbed leg of the plastic versorium. This time we observe that the rubbed plastic versorium leg rotates to point toward the approaching object, as in Figure 3.18.

This experiment shows the opposite of Experiment 3.1. Previously, a rubbed piece of plastic oriented a metal versorium. Now we find that a finger, a piece of wood, or metal orients a rubbed plastic versorium.

In some cases a plastic versorium which has not been intentionally rubbed is attracted by a finger or sheet of paper. As we mentioned before, this is due to the fact that sometimes just handling the plastic versorium is enough to charge it electrically. If this happens, it is an indication that the plastic versorium is not really neutral but rather it has acquired some residual electric charge by being handled. Plastic versoria that are really neutral are versoria that are not attracted nor oriented by these objects.

Experiment 3.9

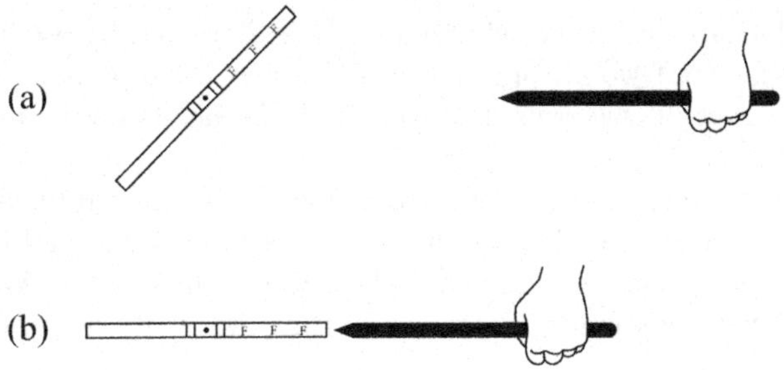

Figure 3.18: (a) The plastic versorium points in an arbitrary direction when it is far away from a wood skewer. (b) The rubbed leg of a plastic versorium is oriented toward a wood skewer which is placed near it.

In Experiment 3.8 we observed the orientation of a plastic versorium, but its center remained at rest on the support. The best way to see a rubbed plastic body being attracted by a metal is to work with the third kind of versorium (see Section 3.1), but now made of plastic, i.e., a plastic strip suspended from its center by a silk or nylon thread. The lighter the strip, the more easily it will move as a whole. On the other hand, it cannot be too short, as we need to rub it on our hair, in a sheet of paper, or a cloth. A plastic straw works fine as a versorium of this kind due to its long length and low weight. If we bring our hand, a sheet of paper, or a metal plate near this neutral versorium, nothing happens to it.

We now rub half of the plastic versorium (that is, we rub only one of its legs) and set it hanging by the silk or nylon thread. Once more we bring our hand near the versorium. We can also bring a sheet of paper, a barbecue wood skewer, or a metal plate near the versorium. This time rubbed plastic versorium not only turns toward this approaching object, but also moves as a whole toward it. That is, both a net torque and a net force are acting upon it. This causes it to be attracted by the approaching object.

This experiment is the opposite of Experiments 2.1, 2.3, and 2.4. We now have a sheet of paper or a metal plate attracting a rubbed piece of plastic. In the case of Experiments 2.1, 2.3, and 2.4, it was not possible to observe or detect this mutual action due to the fact that the plastic straw, ruler, or comb was much heavier than the pieces of paper and of other small objects. Moreover, a person's hand is not sensitive enough to detect the small force exerted by the pieces of paper upon the plastic straw or comb. In Experiments 3.8 and 3.9, on the other hand, we have an instrument with a much greater sensitivity. This is due to the fact that the weight of the test body (in this case the plastic strip of the versorium) is counterbalanced by the support below or above it. As the weight of the plastic versorium has been balanced by another force, it is then much easier to see or detect its rotation or lateral motion due to an external influence.

Experiment 3.10

The fact that an electrically charged body is attracted by other bodies near it (a finger, a piece of wood, or a piece of metal), was used by Stephen Gray (1666-1736) in 1720 to discover new electric materials (that is, materials that behave like amber).

Gray rubbed several substances by running them through his fingers. After this procedure, he checked whether the substances were attracted by a finger or another solid body when brought near the rubbed substances. We quote here sections of his work describing experiments that can easily be reproduced:[13]

> Having often observed in the electrical experiments made with a [rubbed] glass tube, and a down feather tied to the end of a small stick, that after its fibres had been drawn towards the tube, when that [the tube] has been withdrawn, most of them would be drawn to the stick, as if it [the down feather] had been an electric body, or as if there had been some electricity communicated to the stick or feather; this put me upon thinking, whether if a feather were drawn through my fingers, it might not produce the same effect, by acquiring some degree of electricity. This succeeded accordingly upon my first trial, the small downy fibres of the feather next the quill being drawn by my finger when held near it: [...] I then proceeded to try whether hair might not have the same property, by taking one from my wig, and drawing it 3 or 4 times through my fingers, or rather between my thumb and forefinger, and soon found it would come to my finger at the distance of half an inch [1.3 cm]; [...].
>
> Having succeeded so well in these [experiments], I proceeded to larger quantities of the same materials, as pieces of ribband both of coarse and fine silk of several colours, and found that by taking a piece of either of these of about half a yard long [45 cm], and by holding the end in one hand, and drawing it through my other hand between my thumb and fingers, it would acquire an electricity, so that if the hand were held near the lower end of it, it would be attracted by it at the distance of 5 or 6 inches [13 or 15 cm]; but at some times the electricity would be much weaker than at others, the reason of which I conjectured to be, that the ribband might have imbibed some aqueous particles from the moist air, which I found to be [true] upon trial the occasion of it; for when I had well warmed the ribband by the fire, it never failed to be strongly electrical.

3.6 Fabri and Boyle Discover Mutual Electrical Action

Experiments 3.6 to 3.10 are very important. They show that there is a mutual action between the rubbed plastic and the objects around it. The rubbed plastic

[13] [Grab, pp. 104-106].

attracts these objects and they in turn, attract the rubbed plastic. Gilbert did not perform any experiments with a rubbed versorium made of amber or any other electrical substance (i.e., of any substance that behaved like amber, as he called them). This fact, perhaps, contributed to his erroneous belief that there was no mutual action between rubbed amber and surrounding objects. The same erroneous point of view was adopted by Girolamo Cardano (1501-1576) before Gilbert, and was also mentioned by N. Cabeo (1596-1650) after Gilbert. However, they were all aware of the mutual action between two magnets, or between a magnet and a piece of iron. In order to characterize this mutual magnetic action, Gilbert adopted the names *coition* or *confluence*, while for the electrical action he used the term *attraction*.[14]

We now know that electrical action is also mutual, and so it can be characterized by the expression *electrical interaction*. Electrical interaction refers both to the net force exerted by one object on another (causing them to move relative to the ground), and to the net torque exerted by one object on another (causing them to turn relative to the ground). When we talk of electrical force, it should be kept in mind that not only does object A attract object B; object B attracts object A as well in the opposite direction. Likewise, if object A exerts an electrical torque upon object B; object B will cause an opposite torque upon object A. If object A tends to turn in the clockwise direction due to the influence of object B, then B will tend to turn in the counterclockwise direction due to the influence of A.

The first to discover that the electrical action is a mutual action between the rubbed amber and the objects around it were Honoré Fabri (1607-1688) in 1660 (Figure 3.19), and Robert Boyle (1627-1691) in 1675 (Figure 3.20).[15]

Figure 3.19: Honoré Fabri (1607-1688).

[14][Gil78, pp. 26 and 34] and [Hei99, pp. 174-182].

[15][Hei99, pp. 195-205].

Figure 3.20: Robert Boyle (1627-1691).

Fabri became a corresponding member of the Accademia del Cimento in 1660. Among the Academy's members were G. A. Borelli (1608-1671), Vincenzo Viviani (1622-1703)—who was a disciple of Galileo (1564-1642)—and F. Redi (1626-1697/8). It was founded in 1657 and lasted 10 years. The works of the Academy, called *Saggi* or *Essays*, were published in 1667. The studies of electricity done at this Academy began in 1660. Among its reports we find the following comments:[16]

> It is commonly believed, that *amber* attracts the little bodies to itself; but the action is indeed *mutual*, not more properly belonging to the *amber*, than to the bodies moved, by which also itself is attracted.

According to Heilbron, the academicians ascertained this by suspending a piece of rubbed amber by a thread or placing it on a pivot. The amber then, according to the academicians, "made a little stoop to those little bodies, which likewise *proportionally* presented themselves thereto, and readily obeyed its call." In the same year Magalotti (1637-1712) refuted Cabeo's points of view; Cabeo had rejected the mutual nature of electrical interactions. Magalotti said:[17] "His views are refuted by experience, for the *ambra versoria* follows all bodies presented to it." According to Heilbron, Magalotti obtained this information from Fabri. The previous reports of the Accademia del Cimento were also due to Fabri, according to Heilbron. There is a manuscript containing drafts for a section on electricity of the Essays, in Fabri's letter, mentioning that "a piece of sealing wax suspended freely and then rubbed approaches other bodies."[18] From these statements we can see that these experiments by Fabri were somewhat similar to the ones presented in Section 3.5, in which we used rubbed plastic instead of rubbed amber or rubbed wax.

[16][Hei99, p. 201]
[17][Hei99, p. 201].
[18][Hei99, p. 202].

Boyle presented his results on the mutual action between rubbed amber and nearby objects in 1675. He may have learned of this from Fabri's reports or he may have discovered the fact independently. He believed that amber emitted a material effluvium which would cause it to attract light bodies, perhaps due to the fact that the effluvium might be sticky and elastic.

As regards the attraction exerted by amber, he wrote the following:[19]

> That 'tis not in any peculiar sympathy between an electric and a body whereon it operates, that electrical attraction depends, seems the more probable, because amber, for instance, does not attract onely one determinate sort of bodies, as the loadstone does iron, and those bodies wherein it abounds; but as far as I have yet tried, it draws indifferently all bodies whatsoever, being plac'd within a due distance from it, (as my choicest piece of amber draws not onely sand and mineral powders, but fillings of steel and copper, and beaten gold it self) / provided they be minute or light enough, except perhaps it be fire.

In another passage comes the crucial realization:[20]

> We have found by experiment, that a vigorous and well excited piece of amber will draw, not onely the powder of amber, but less minute fragments of it. And as in many cases one contrary directs to another, so this trial suggested a further, which, in case of good success, would probably argue, that in electrical attraction not onely *effluvia* are emitted by the electrical body, but these *effluvia* fast upon the body to be drawn, and that in such a way, that the intervening viscous strings, which may be supposed to be made up of those cohering *effluvia*, are, when their agitation ceases, contracted or made to shrink inwards towards both ends, almost as a highly stretch'd lute-string does when 'tis permitted to retreat into / shorter dimensions. But the conjecture itself was much more easie to be made than the experiment requisite to examine it. For we found it no easie matter to suspend an electric, great and vigorous enough, in such a manner, that it might, whilst suspended, be excited, and be so nicely poised, that so faint a force as that wherewith it attracts light bodies should be able to procure a local motion to the whole body it self. But after some fruitless attempts with other electricks, I had recourse to the very vigorous piece of polish'd amber, formerly mention'd, and when we had with the help of a little wax suspended it by a silken thread, we chased very well one of the blunt edges of it upon a kind of large pin-cushion cover'd with a course and black woollen stuff, and then brought the electric, as soon as we could, to settle notwithstanding its hanging freely at the bottom of the string.

[19][Boy00, p. 515].

[20][Boy00, p. 516].

> This course of rubbing on the edge of the amber we pitch'd / upon for more than one reason; for if we had chased the flat side, the amber could not have approached the body it had been rub'd on without making a change of place in the whole electric, and, which is worse, without making it move (contrary to the nature of heavy bodies) somewhat upwards; whereas the amber having, by reason of its suspension, in parts counterpoised by one another; to make the excited edge approach to another body, that edge needed not at all ascend, but onely be moved horizontally, to which way of moving the gravity of the electric (which the string kept from moving downwards) could be but little or no hinderance. And agreeably to this we found, that if, as soon as the suspended and well rubb'd electric was brought to settle freely, we applied to the chafed edge, but without touching it, the lately mention'd cushion, which, by reason of its rough *superficies* and porosity, was fit for the electrical / *effluvia* to fasten upon, the edge would manifestly be drawn aside by the cushion steadily held, and if this were slowly removed, would follow it a good way; and when this body no longer detain'd it, would return to the posture wherein it had settled before. And this power of approaching the cushion by vertue of the operation of its own steams, was so durable in our vigorous piece of amber, that by once chasing it, I was able to make it follow the cushion no less than ten or eleven times.

The experiments of Fabri and Boyle showed that not only rubbed amber oriented and attracted lightweight objects, but that also the rubbed amber and rubbed sealing wax were oriented and attracted by other objects. Their experiments are diametrically opposite to Fracastoro's observations, described in Section 3.1. Fracastoro suspended small pieces of amber and silver in his perpendiculo and observed that they were attracted by another piece of rubbed amber brought close to them, as in Figure 3.2. Fabri and Boyle, on the other hand, observed that a rubbed amber suspended by a thread was attracted and oriented by another object placed close to it, as in Figure 3.21.

This is a very important physical discovery. It shows that there is action and reaction in electrostatics. That is, there is a mutual electrical interaction between the rubbed object and the nearby objects. The rubbed object exerts a force and a torque on nearby neutral objects. And these objects in turn exert an opposite force and an opposite torque on the rubbed object.

In 1660 and 1675 Fabri and Boyle concluded experimentally that there was action and reaction in electricity. These were only qualitative proofs, like the experiments described in this Section. They did not measure the force exerted by the amber or the opposite force exerted by the surrounding bodies.

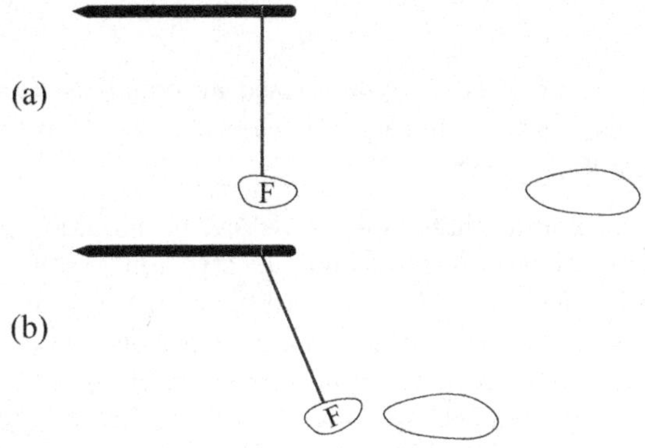

Figure 3.21: (a) A rubbed piece of amber, represented by F, hangs vertically when it is far away from any neutral object. (b) The rubbed piece of amber F being attracted by a neutral body brought close to it.

3.7 Newton and Electricity

As we saw in Section 3.6, between 1660 and 1675 Fabri and Boyle discovered that electrical actions are mutual. A few years later, in 1687, Isaac Newton (1642-1727) (Figure 3.22), included the principle of action and reaction as one of the pillars of the whole of physics.

Figure 3.22: Isaac Newton (1642-1727). This is the most famous portrait of Newton. It was made by Godfrey Kneller (1646-1723) in 1689. Newton appears with his natural hair, at the peak of his scientific career, two years after the publication of the *Principia*.

This is the third axiom or law of motion which he included in his famous book *Mathematical Principles of Natural Philosophy*, also known by the Latin name, *Principia*. His third axiom, or third law of motion, was formulated as follows:[21]

> To every action there is always opposed an equal reaction: or, the mutual actions of two bodies upon each other are always equal, and directed to contrary parts.

It is important to note that Newton believed this axiom should be valid for all known interactions: gravitational, electric, magnetic, elastic, contact forces, friction, collisions, etc. Moreover, to Newton this principle had both a qualitative meaning and a very precise quantitative aspect, namely, that to every action there is always a reaction of equal intensity. Moreover, they are aligned along the straight line connecting the two bodies, directed to corresponding parts. By action and reaction Newton referred to the mutual forces between the two bodies, which are measured quantitatively by variations of the linear momentum of each body per unit time. What we call linear momentum Newton called quantity of motion, since it is proportional to the mass of the body multiplied by its velocity in relation to absolute space. This fundamental work by Newton marks a new epoch in physics, an epoch in which science attained high degree of precision.

In the *Principia* Newton presented collisions and contact forces as examples of interactions that satisfy the principle of action and reaction. As regards actions at a distance, he quoted the examples of gravitation and magnetism, discussing the latter in the following words:[22]

> In attractions, I briefly demonstrate the thing after this manner. Suppose an obstacle is interposed to hinder the meeting of any two bodies A, B, attracting one the other; then if either body, as A, is more attracted towards the other body B than that other body B is towards the first body A, the obstacle will be more strongly urged by the pressure of the body A than by the pressure of the body B, and therefore will not remain in equilibrium: but the stronger pressure will prevail, and will make the system of the two bodies, together with the obstacle, to move directly towards the parts on which B lies; and in free spaces, to go forwards *in infinitum* with a motion continually accelerated; which is absurd and contrary to the first Law. For, by the first Law, the system ought to continue in its state of rest, or of moving uniformly forwards in a right line; and therefore the bodies must equally press the obstacle, and be equally attracted one by the other. I made the experiment on the loadstone and iron. If these, placed apart in proper vessels, are made to float by one another in standing water, neither of them will propel the other; but, by being equally attracted, they will sustain each other's pressure, and rest at last in an equilibrium.

In Query 8 of his book *Optics*, Newton described experiments on electricity:[23]

---

[21] [New52a, p. 14].

[22] [New52a, p. 22].

[23] [New52b, pp. 516-517].

> Qu. 8. Do not all fixed bodies, when heated beyond a certain degree, emit light and shine; and is not this emission performed by the vibrating motions of their parts? And do not all bodies which abound with terrestrial parts, and especially with sulphureous ones, emit light as often as those parts are sufficiently agitated; whether that agitation be made by heat, or by friction, or percussion, or putrefaction, or by any vital motion, or any other cause? [...] So also a globe of glass about 8 or 10 inches in diameter, being put into a frame where it may be swiftly turned round its axis, will in turning shine where it rubs against the palm of one's hand applied to it. And if at the same time a piece of white paper or white cloth, or the end of one's finger be held at the distance of about a quarter of an inch or half an inch from that part of the glass where it is most in motion, the electric vapour which is excited by the friction of the glass against the hand will (by dashing against the white paper, cloth or finger) be put into such an agitation as to emit light, and make the white paper, cloth, or finger appear lucid like a glow-worm; and in rushing out of the glass will sometimes push against the finger so as to be felt. And the same things have been found by rubbing a long and large cylinder or glass or amber with a paper held in one's hand, and continuing the friction till the glass grew warm.

Electricity is also mentioned in Query 31, where Newton emphasizes the mutual interactions at a distance between the bodies:[24]

> Qu. 31. Have not the small particles of bodies certain powers, virtues, or forces, by which they act at a distance, not only upon the rays of light for reflecting, refracting, and inflecting them, but also upon one another for producing a great part of the phenomena of Nature? For it's well known that bodies act one upon another by the attractions of gravity, magnetism, and electricity; and these instances shew the tenor and course of Nature, and make it not improbable but that there may be more attractive powers than these. For Nature is very consonant and conformable to herself. How these attractions may be performed I do not here consider. What I call attraction may be performed by impulse, or by some other means unknown to me. I use that word here to signify only in general any force by which bodies tend towards one another, whatsoever be the cause. For we must learn from the phenomena of Nature what bodies attract one another, and what are the laws and properties of the attraction, before we enquire the cause by which the attraction is performed. The attractions of gravity, magnetism, and electricity reach to very sensible distances, and so have been observed by vulgar eyes, and there may be others which reach to so small distances as

[24] [New52b, pp. 531].

> hitherto escape observation; and perhaps electrical attraction may reach to such small distances, even without being excited by friction.

He also mentioned electrical attractions in the *Principia.* For instance, in Book III:[25]

> Proposition 7. Theorem 7
>
> *That there is a power of gravity pertaining to all bodies, proportional to the several quantities of matter which they contain.*
>
> [...]
>
> Cor. I. Therefore the force of gravity towards any whole planet arises from, and is compounded of, the forces of gravity towards all its parts. Magnetic and electric attractions afford us examples of this; for all attractions towards the whole arises from the attractions towards the several parts. [...]

In the General Scholium at the end of the book he also mentioned electricity, emphasizing again the mutual interactions between bodies:[26]

> And now we might add something concerning a certain most subtle spirit which pervades and lies hid in all gross bodies; by the force and action of which spirit the particles of bodies attract one another at near distances, and cohere, if contiguous; and electric bodies operate to greater distances, as well repelling as attracting the neighboring corpuscles; and light is emitted, reflected, refracted, inflected, and heats bodies; and all sensations is excited, and the members of animal bodies move at the command of the will, namely, by the vibrations of this spirit, mutually propagated along the solid filaments of the nerves, from the outward organs of the sense to the brain, and from the brain into the muscles. But these are things that cannot be explained in few words, nor are we furnished with that sufficiency of experiments which is required to an accurate determination and demonstration of the laws by which this electric and elastic spirit operates.

These quotations show that, in Newton's mind, electricity played a fundamental role in natural phenomena.

[25] [New52a, pp. 281-282].

[26] [New52a, p. 372].

Chapter 4

Electrical Attraction and Repulsion

4.1 Is There Electrical Repulsion?

All experiments described thus far in this book have dealt with the attraction between bodies. But electrical phenomena are also characterized by repulsion.

Experiment 4.1

A very simple way to observe electrical repulsion is with a strip from a plastic bag. Cut a 2 cm wide by 10 to 20 cm long strip. Hang it over a horizontal support (a wood skewer, pencil, or finger). The two halves of the neutral strip initially hang vertically parallel to each other, as in Figure 4.1 (a). We now rub both halves with the same material (for instance, by passing each half through our fingers, or rubbing them in a piece of paper). After release they repel one another, with the two halves moving away laterally (Figure 4.1 (b)).

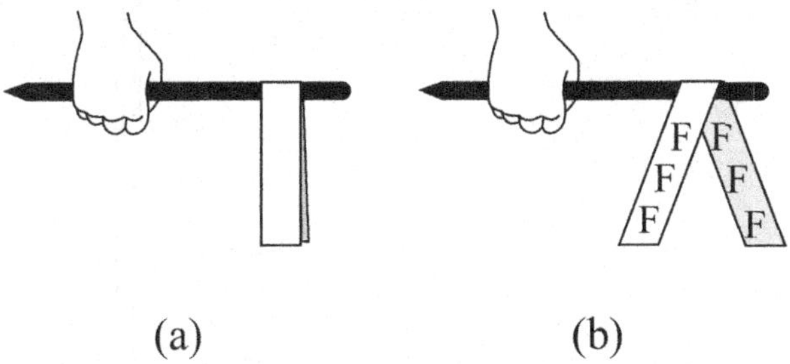

Figure 4.1: (a) A neutral plastic strip hangs vertically from a horizontal wood skewer. (b) Repulsion between the two halves of a rubbed plastic strip.

It is also possible to see this effect with two pieces of a plastic straw, each around 5 cm long. One end of each straw is tied to each end of a silk or nylon thread 10 or 20 cm long. The thread hangs from its center on a horizontal support in such a way that both neutral straws remain initially side by side vertically, with their free ends pointing downward. When the two straws are rubbed with the same material, they begin to repel one another after release, moving outward. The thinner the horizontal support, the more visible the phenomenon will be.

A third alternative utilizes inflated rubber balloons. Two balloons, supported by threads, are initially hung such that they are touching one another. When the balloons are rubbed on our hair, they begin to repel one another. They move away from one another and are no longer touching.

In all these cases we observe the repulsion of bodies made of the same material (two plastic strips, two straws, or two rubber balloons) which were rubbed with the same substance. This is a new phenomenon which was not seen in the previous experiments.

Experiment 4.2

By utilizing two plastic versoria we can observe electrical orientation due to repulsion. We rub only one leg of each versorium with the same substance—for instance, a sheet of paper or our hair. The two versoria are placed side by side, parallel to one another, with the rubbed legs held pointing in the same direction. After release from rest the rubbed legs repel one another. As a result of this repulsion, the versoria rotate relative to their vertical axes in such a way that the end of the process the rubbed legs remain aligned but at the greatest possible distance from one another (Figure 4.2).

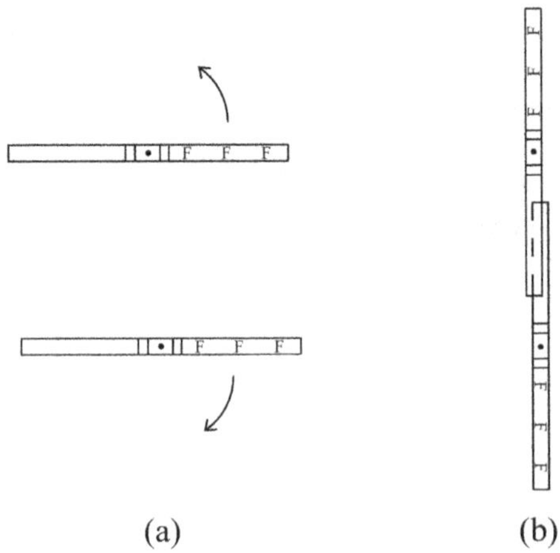

Figure 4.2: (a) Repulsion between the rubbed legs of two plastic versoria, as viewed from above, placed side by side along parallel directions. (b) Final equilibrium orientation of the two versoria.

This effect is more visible if the two versoria are initially very close to one another. In order to prevent contact between the two legs which have not been rubbed after the versoria have rotated, you can place one of the versoria slightly higher than the other in such a way that after the repulsion takes effect, the leg of one versorium will be above the leg of the other.

It should also be observed that the versoria rotate in opposite directions after release. While one rotates clockwise, the other rotates counterclockwise. This means that the torque exerted by versorium A upon versorium B is opposite to the torque exerted by versorium B upon versorium A.

In order to remember more easily which leg has been rubbed, we can mark it with a drop of ink, a pen, or a small cut.

Experiment 4.3

Another variation on Experiment 4.2 is to rub both legs of each versorium. They are then placed aligned in a straight line with one end of each very close to one another, almost touching. After release from rest they rotate in opposite directions. When they stop, they are parallel to one another, side by side, as in Figure 4.3.

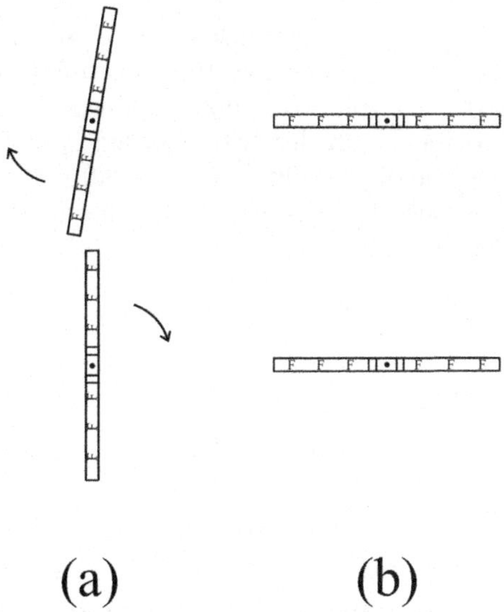

Figure 4.3: (a) Repulsion between two plastic versoria, as viewed from above, which have been rubbed along their entire lengths when they are placed close to one another and aligned in the same direction. (b) Final equilibrium orientations of the two versoria.

4.2 Guericke's Experiment with a Floating Down Feather

Experiment 4.4

We now perform a new experiment, analogous to one which had great historical importance. Initially we take an object like a down feather, a dandelion seed, or a few strands of cotton. The important factor is that the object selected should take a long time to fall to the ground in air, e.g., some 10 seconds to cover a distance of 2 meters of fall. It is even better if it falls slower than this. On the other hand, if it falls much faster than this, it will not be possible to observe the effects described here.

We begin by rubbing on hair a plastic straw. In order to determine whether the straw is well charged we can use the wall test described in Experiment 3.6. The rubbed straw is then held horizontally in our fingers at one end. With the other hand we release the feather, dandelion seed, or piece of cotton a little above the straw. The object is attracted by the straw and sticks to it. If we look closely at the object we can see its strands stretching, as if they wanted to move away from the straw. This occurs for the same reason the objects used in Experiment 4.1 repel each other. Sometimes the object actually jumps upward after contact, moving away from the rubbed straw. If this does not happen immediately, we can induce the object to release by tapping on the straw, or by blowing the object softly. After the object is free from the straw and begins falling, we can place the rubbed straw below the falling object. The object is then repelled by the straw and moves upward. Sometimes this does not happen at once, since the object must touch the rubbed straw two or three times and be freed after each touch before it can clearly be repelled by it. The more electrified the straw, more quickly the object will be repelled after touching it. From now on we will suppose that the object is already floating in air, being repelled by the rubbed straw below it, as in Figure 4.4.

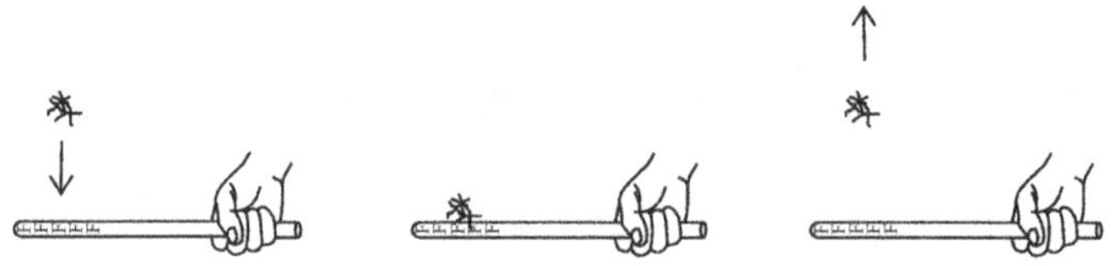

(a) (b) (c)

Figure 4.4: (a) A piece of cotton is initially attracted by a rubbed plastic straw. (b) The cotton touches the rubbed portion of the straw. (c) After contact, the cotton is repelled by the straw. It can then be kept floating above the straw despite the gravitational attraction of the Earth!

Figure 4.5 illustrates a similar experiment made with a dandelion seed. The dandelion seed falls naturally very slowly, so it is suitable for this experiment.

It is easy to keep it floating above a plastic straw rubbed with hair.

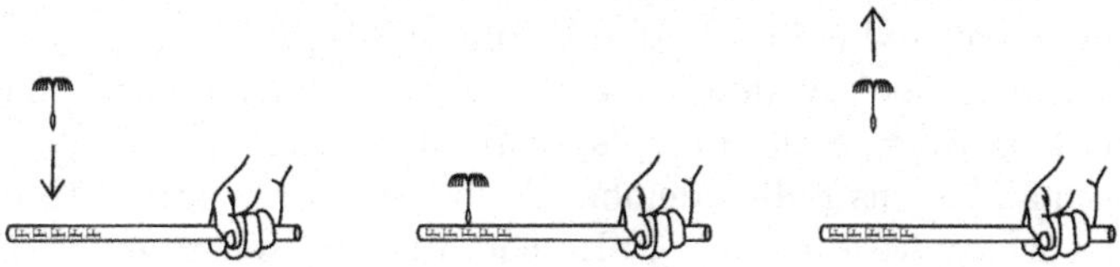

(a) (b) (c)

Figure 4.5: Experiment 4.4 can easily be performed with a dandelion seed floating above a plastic straw rubbed with hair.

By moving the rubbed straw slowly below the object, we can move it to any place inside a room. If the object comes very close to our body or to any other item in the room, it is attracted to our body or item and sticks to it. To prevent this from happening, we utilize the rubbed straw to propel the object away from these bodies. In this case the object can easily be kept floating for some time at a distance of 10 to 20 cm above the straw, depending upon how well electrified the straw is. To keep the object floating the rubbed straw must be kept moving constantly below it, following the motion of the object, in order to guide its motion. With a dandelion seed the procedure is normally easier. When it is first released in the air above the rubbed straw, it is attracted by the straw, touches it and is immediately repelled by it. This is a very simple experiment, but extremely curious. No one who performs it forgets what he or she sees.

An experiment like this was of tremendous historical importance. It was performed for the first time by Otto von Guericke (1602-1686) (Figure 4.6).[1]

Figure 4.6: Otto von Guericke (1602-1686).

[1][Hei99, pp. 215-218].

It appeared in his book *The New (So-Called) Magdeburg Experiments*, published in 1672, in Latin. According to the preface of this work, the book had actually been completed in 1663. At one point in his life Guericke was mayor of Magdeburg. In this book he described the air pump machine which he invented based on the discovery of the pumping capacity of air. In 1657 he used it to perform the now famous public demonstration at Magdeburg of the enormous forces due to atmospheric pressure.[2] He had a hollow sphere made of two copper hemispheres which touched one another side by side. The air inside the sphere was removed with his pump. After the air was removed, two groups of eight horses pulling on each side had enormous difficulty separating the hemispheres. On the other hand, when air was allowed to enter the sphere, the two hemispheres could be easily separated by one person.

But what interests us here is another experiment performed by Guericke. His illustration describing this experiment is presented in Figure 4.7.

Figure 4.7: Guericke's experiment in which he kept a feather floating above a rubbed sulphur globe, together with his electrical machine.

We quote here from his famous book:[3]

> *The Experiment Wherein these Aforementioned Important Virtues can be Excited through Rubbing on a Sulphur Globe*
>
> If one is so minded, he should take a glass sphere, a so-called phial, the size of a baby's head and pour in sulphur that has been powered in a mortar. Then, by heating it, he should cause the powder to melt. After cooling it he should break the glass sphere, extract the ball which remains and store it away in a dry place of low humidity.
>
> [...]
>
> *Section 2.*

[2][Kra81].

[3][Gue94, Book 4, Chapter 15, pp. 227-231].

> In order to demonstrate the conserving virtue present in this globe, one should set it up with a rod through its core on two supports, *ab*, on a stand labelled, *abcd*. This should be a palm's width in height from the base and all kinds of shreds or bits of leaves, gold, silver, paper, hop plants and other tiny particles should be strewn beneath it. Then one should touch the sphere with a dry hand and rub or stroke it two or three times, etc. At this point it will attract the aforementioned fragments to it. If the globe is rotated on its axis, it will carry these bits along with it. Now we can visually perceive how the sphere of our Earth holds and maintains all animals and other bodies on its surface and carries them about with it in its daily twenty-four hour motion.
>
> [...]
>
> *Section 3.*
>
> One can clearly demonstrate the presence of the expulsive virtue in this globe when it is removed from the aforementioned stand and being held in the hand, is rubbed or stroked in the manner already described. Then it not only attracts, but also again repels from itself small bodies of the kind mentioned above (depending upon the prevailing weather). Once it has touched these bodies, it does not attract them again until they have subsequently touched some other body. This virtue can be seen particularly clearly in its effect upon very soft and light feather, *a*, (because they fall to Earth more slowly than other bits and shreds). Thus when the feathers are propelled upwards and hang in the sphere of virtue of this globe, they can float for quite a long time and be carried around the whole room with the globe wherever one wishes.
>
> [...]

Experiment 4.4 is analogous to Guericke's experiment, but uses a plastic straw instead of a sulphur globe. However, it must be emphasized that Guericke himself did not consider the repulsion of the feather or, as he put it, the "expulsive virtue" of the globe, as an intrinsically electrical phenomenon. To Guericke this property of the sulphur globe was analogous to the repulsive power shown by the Earth in some circumstances. As a result, it is now felt that Guericke neither discovered nor recognized a genuine electrical repulsion.

Guericke's apparatus is regarded by some authors as the first electrical machine, i.e., the first equipment with which objects could be electrified. But Guericke himself would probably not agree with this statement. For Guericke, the sulphur ball was a simulacrum of the Earth. The several "virtues" shown by the ball, attractive and repulsive, would simulate the analogous virtues of our planet. Therefore, for Guericke these virtues were not something genuinely electrical. A detailed analysis of this subject can be found in the works of Roller and Roller, Krafft, and Heilbron.[4]

The first instrument built intentionally to produce the electrification of bodies is due to Hauksbee (approximately 1666-1713) (Figure 4.8).[5] The crank and pulley turn the small wheel, which turns the glass globe. Hauksbee rubbed the external surface of the rotating globe with a sheet of paper or with his bare hands. An instrument like this is called an *electrical machine*, an *electrostatic machine*, a *triboelectric generator*, or a *frictional electric generator.*

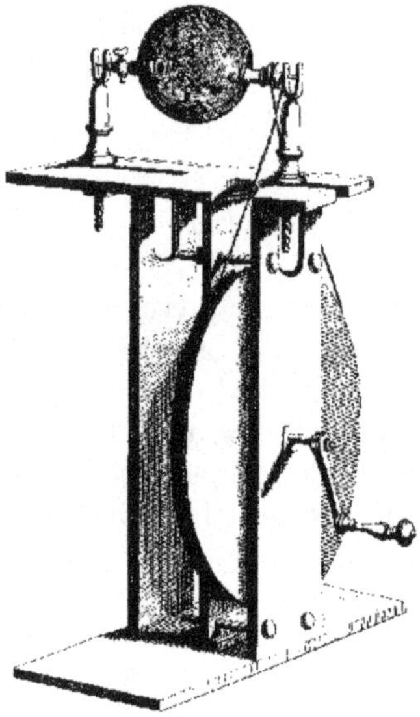

Figure 4.8: Hauksbee's electric machine.

Experiments analogous Guericke's were conducted by Gray and by Francis Hauksbee in 1708. In the experiments of Gray and Hauksbee the feather was attracted and then repelled by a rubbed flint-glass, which is a special kind of glass containing lead in its composition.[6] Later on we will see how this experiment played a crucial role in a great discovery made by Du Fay. Gray's 1708 paper was not published until 1954.[7] Gray does not quote Guericke's book, but it is possible that he knew his work, although this is not known for sure. Hauksbee saw Gray's original paper and had a role in suppressing its publication.[8] Hauksbee published analogous experiments with floating feathers without mentioning Gray's name.

[4][RR57, pp. 565-568], [Kra81], and [Hei99, pp. 215-216].

[5][Hau09, Plate VII], [RR57, pp. 565-568], [Hom67], [Hom81, pp. xiv-xv, 14, 42, 77, and 78n], [Que], [Hei81d], and [Hei99, pp. 230-234].

[6][Chi54], [Haub], [RR57, pp. 570 and 584-585], [Hom81, p. 13], and [Hei99, pp. 235-236].

[7][Chi54].

[8][Hei81c] and [Hei99, p. 236].

In his paper of 1708 Gray described twelve experiments using a glass tube 70 to 80 cm long with a diameter of 2 to 3 cm. He rubbed the tube with his bare hands. Here we quote only the four initial experiments:[9]

> Exper. 1st A down feather being let goe from the fingers came to the [rubbed] glass [tube] at the distance of more than 30 inches [76 cm] some of the smallest fibres answered to the motion of then hand whilst the glass was rubing at the distance of more than 50 inches [1.3 m]. [An illustration of this experiment appears in Figures 4.9 and 4.10.]

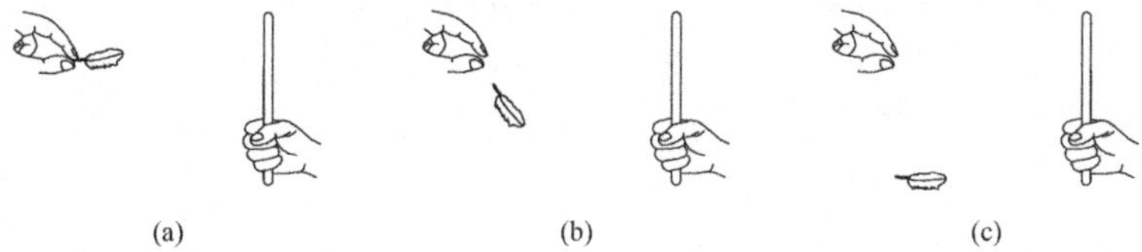

Figure 4.9: A down feather is released close to an unrubbed glass tube. If falls to the ground.

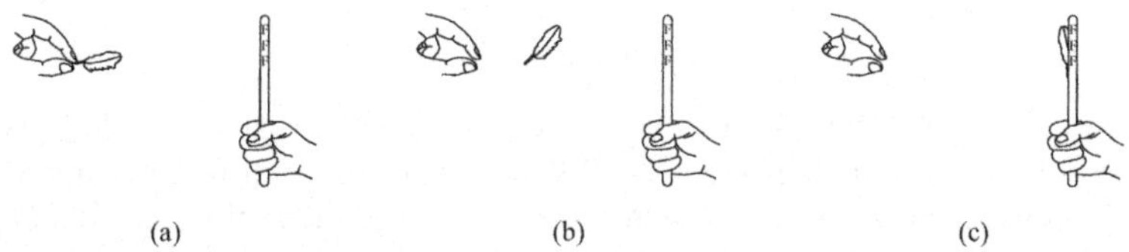

Figure 4.10: Gray's experiment showing a down feather being attracted by a rubbed glass.

> Exp. 2d if when the feather is come to the glass it be held at about 6 or 8 inches [15 or 20 cm] distance from the side of a wall edge of a table arme of a chair or the like it will be drawn to it and thence to the glass again and that for 10 or 15 times together without ceasing it flies to object at a greater distance but then does not soe often return. [An illustration of this experiment appears in Figure 4.11.]

> Exp. 3d when the feather is on the glass and half of its fibres are extended towards it the other from it diverging in two cones that remotest from the glass is much more obtuce than the other if when the feather is in this posture you pinch its fibres between your thumbe and finger they will draw back soe soon as let goe and imediatly

[9][Chi54, pp. 34-35].

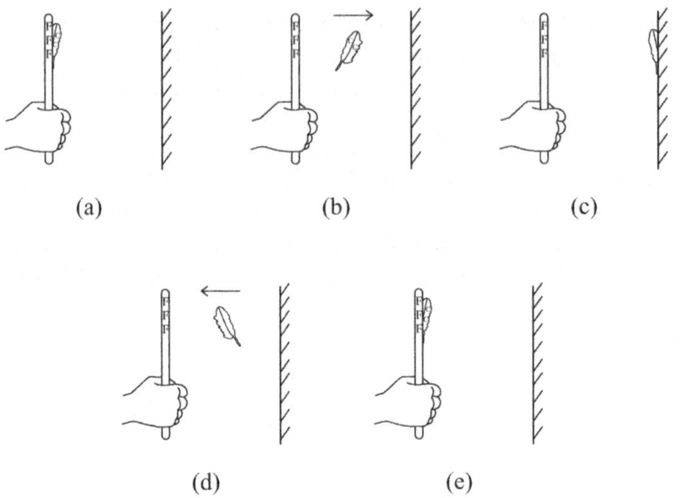

Figure 4.11: Gray's experiment showing a down feather oscillating between a rubbed glass and a wall.

> cleave to the glass and as if they had retained some sence of the injurie offerd them will hardly be allured to salute your fingers again but this is not ollwais alike precented.
>
> Exp. 4th When the feather is come to the glass and thence reflected if you follow it with the glass twill flee from it and will by noe means be made to touch it till driven near to the next wall in the room or some other solid object by which twill be attracted and freely return to the glass again repeating its reflections as in the second experiment soe I have sometimes caried the feather round the room at the distance of 5 or 6 inches [13 or 15 cm] without touching it and could move it upwards downwards inclineing or horizontally in a line or circle according to the motion of the glass and if when the feather was floating in the air I rubed the glass the feather would remove farther from it yet would respond to the motion of my hand by a vibrating motion not to be accounted for by that of the air.

4.3 Du Fay Recognizes Electrical Repulsion as a Real Phenomenon

The experiments described in these Sections bring something new and extremely important. Up to now we have only observed attraction or the absence of attraction between a rubbed body and the surrounding light substances. Now we saw that there is also electrical *repulsion*.

Although electrical repulsion had been observed a few times by some researchers, it was normally interpreted as a secondary effect. Sometimes this phenomenon was interpreted as only an *apparent* repulsion. We present here a few alternative explanations. (a) Some people believed that this behaviour was due to an air flux that moved the light substances away from the rubbed body. (b) Another interpretation was that this *apparent* repulsion was in fact an attraction exerted by other surrounding bodies upon the light substance. That is, according to this interpretation, it was not the rubbed amber which began to repel the light substance, but the light substance was actually being attracted by other surrounding bodies which had somehow acquired some electricity. If this were the case, the light substance might move away from the rubbed amber if it experienced a smaller attraction from the amber than from the surrounding bodies. (c) Another interpretation sometimes given for the observed repulsion was that it is like a collision. That is, the body would initially be attracted by the rubbed amber, collide with it, and then rebound to a distance. This rebound was interpreted mechanically, not as a genuine electrical repulsion.

It was only with the publication of the works by Charles François de Cisternay Du Fay in 1733 and 1734 that the repulsion was recognized as a legitimate and characteristic phenomenon of electrical interactions.[10] There is a nice biography of Du Fay (Figure 4.12), written by Heilbron.[11] Following the initial works of Gray, Du Fay succeeded in publishing some remarkable papers with fundamental discoveries on electricity.[12]

Figure 4.12: Du Fay (1698-1739).

Here we quote Du Fay's words describing how he concluded that electrical repulsion was a real phenomenon (our emphasis in italics).[13] It is interesting

[10] [Hei99, pp. 5 and 255-258].
[11] [Hei81b].
[12] [DF33a], [DF33c], [DF33d], [DF33b], [DF], [DF34a], [DF34b], [DF37b], and [DF37a].
[13] [DF33b, pp. 457-458].

to note that Du Fay himself did not initially consider repulsion to be a real phenomenon, and later changed his mind due to the experimental evidence:[14]

> *On the Attraction and Repulsion of Electric Bodies.*
>
> Until today we always considered the electric virtue in general, and by this word we understand not only the property electric bodies have of attraction, but also the property of repelling the bodies which they had attracted. This repulsion is not always constant, and it is subject to variations which made me consider the subject attentively, and I believe to have discovered some very simple principles which had not yet been suspected, and which explain all these variations, in such a way that I do not know up to now any experience which is not in agreement [with these principles] in a very natural way.
>
> I observed that light bodies are normally only repelled by the [rubbed glass] tube when we approach [these light bodies] with any [other] bodies having an appreciable volume, and this made me think that these last [large] bodies had been electrified by the approach of the tube and that in this way it [the light body] was always attracted, either by the tube, or by the surrounding [large] bodies, *in such a way that there was never a real repulsion.*
>
> However, an experiment indicated to me by Mr. de Reaumur [René Antoine Ferchault de Réaumur, (1683-1757)] opposed itself to this explanation. It consists in placing gun powder near the edge of a card, approaching this powder with a tube of electrified Spanish

---

[14] "*De l'Attraction & Répulsion des Corps Électriques.*

Nous avons toûjours considéré jusqu'a présent la vertu électrique en général, & sous ce mot on a entendu non seulement la vertu qu'ont les corps électriques d'attirer, mais aussi celle de repousser les corps qu'ils ont attirés. Cette repulsion n'est pas toûjours constante, & elle est sujette à des variétés qui m'ont engagé à l'examiner avec soin, & je crois avoir découvert quelques principles très-simples qu'on n'avoit point encore supçonnés, & qui rendent raison de toutes ces varietés, ensorte que je ne connois jusqu'à présent aucune expérience qui ne s'y accorde très-naturellement.

J'avois observé que les corps légers n'etoient ordinairement repoussés par le tube que lorsque l'on en approchoit quelque corps d'un volume un peu considérable, & cela me faisoit penser que ces derniers corps étoient rendus électriques par l'approche du tube, & qu'alors ils attiroient à leur tour le duvet, ou la feuille d'or, & qu'ainsi il étoit toûjours attiré, soit par le tube, soit par les corps voisins, mais qu'il n'y avoit jamais de répulsion réelle.

Une expérience que M. de Reaumur m'indiqua, s'oppsoit à cette explication; elle consiste à poser au bord d'une carte un petit monceau de poudre à mettre sur l'écriture, on approche de ce monceau un bâton de cire d'Espagne rendu électrique, & on voit très-clairement qu'il chasse au de-là de la carte des particules de poudre, sans qu'on puisse supçonner qu'elles soient attirées par aucun corps voisin.

Une autre expérience aussi simple, & encore plus sensible, *acheva de me prouver que ma conjecture étoit fausse.* Si l'on met des feuilles d'or sur une glace, & que l'on approche le tube par dessous, les feuilles sont chassées en haut sans retomber sur le glace, & on ne peut certainement expliquer ce mouvement par l'attraction d'aucun corps voisin. La même chose arrive à travers la gaze de couleur, & les autres corps qui laissent passer les écoulements électriques, en forte qu'on ne peut pas douter qu'il n'y ait une répulsion réelle dans l'action des corps électriques."

wax, and observing that it draws back any grains beyond the card. In this case it cannot be suspected that the grains were attracted by a neighboring body.

Another experiment as simple as this one, and even more sensitive, *convinced me that my conjecture was false.* By placing gold leaves onto a crystal, by approaching the [rubbed glass] tube from below, the gold leaves are expelled upwards and do not fall to the crystal. Certainly we cannot explain this [upward] motion by the attraction of a neighboring body. The same thing happens through a colored gauze and through other bodies which allow the passage of the electric flows, in such a way that we cannot doubt that a real repulsion does not exist in the action of the electric bodies.

4.4 The Electric Pendulum

To detect other electric phenomena more clearly we need some specific instruments. We will now build an *electric pendulum*, also called an *electrostatic pendulum.* The simplest procedure is to tie a silk thread to a horizontal support, like a plastic straw. You can also use a thin synthetic polyamide thread, such as nylon, or polyester thread. *The important point is that this thread should not be made of cotton or linen, like a sewing thread.* We tie a piece of paper or aluminum foil to the lower free end of the thread. This piece of paper can be a disk, a square, a triangle, etc., with a diameter or greatest length on the order of 1 or 2 cm. For the time being its shape will not be relevant. The piece of paper should not be crumpled or attached with adhesive tape. Adhesive tape can prevent some phenomena that will be described here from being observed. It is best to make a hole in the piece of paper with a pin, fastening the thread though it (Figure 4.13). Normally aluminum foil works better than normal paper. In Section 6.5 we will introduce the fundamental components of an electric pendulum like this one, after we have performed several experiments with it.

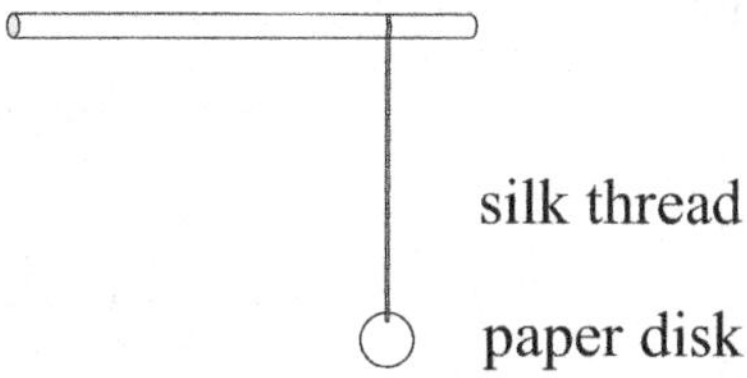

Figure 4.13: A simple way to make an electric pendulum.

Another practical method relies on fixed plastic straws. Initially we make a support for the whole system. It can be a piece of modeling clay with a nail or paper fastener stuck through it. The nail or paper fastener will be located inside the straw, supporting it in a vertical position, so the nail's thickness and length should be chosen appropriately.

A very practical support can be made with thin plastic coffee cups. We make a small hole at the bottom of the cup and push both legs of a paper fastener pass through it. The cup is placed with its mouth upward. We fill it with wet gypsum dough or wet white cement. It will dry in this position. It will be used with the cup's mouth facing downward and the paper fastener pointing upward (Figure 4.14). Because this kind of support will also be used in other electrical instruments, it is useful to prepare several of these at once. Some experiments may require as many as 10 support bases.

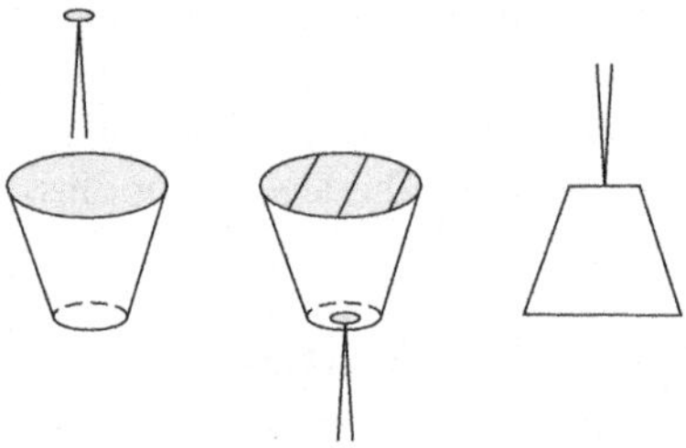

Figure 4.14: Support for the electric pendulum made of a thin plastic coffee cup, paper fastener, and gypsum dough.

After this procedure, we place a drinking straw in the shape of the upside down letter *L* upon the support. Another alternative is to use two straws, one set vertically on the support and the other horizontal. The second straw is attached to the first straw with a second paper fastener. The legs of this second paper fastener have an angle of 90°, with one leg vertical and the other horizontal.

At the free end of the horizontal straw we tie the silk thread with the paper disk attached to its lower end. This completes the electric pendulum (Figure 4.15).

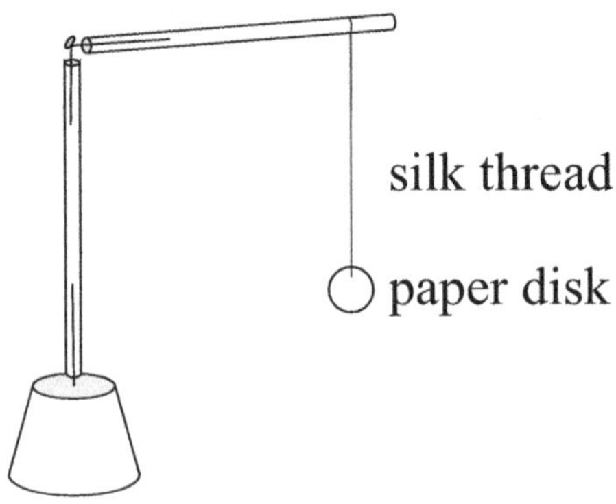

Figure 4.15: Electric pendulum with support.

Experiment 4.5

We mount an electric pendulum with a paper disk at its end. We wait until the pendulum is at rest, with the silk thread and paper disk hanging vertically downward. We take now a neutral plastic straw, that is, a straw which does not attract light pieces of paper on a table, as in Experiment 2.1 (Figure 2.1). We bring this neutral plastic straw near the paper disk of the electric pendulum. Nothing happens. The pendulum remains at rest in the vertical direction.

We rub another straw in a sheet of paper, in hair, or in a tissue. The rubbed straw is slowly brought near the pendulum. We observe that the paper disk begins to move toward the rubbed portion of the straw. For the time being we will not bring them into contact. In this situation the silk thread of the pendulum remains at an angle to the vertical, with the paper disk close to the rubbed section of the plastic straw (Figure 4.16).

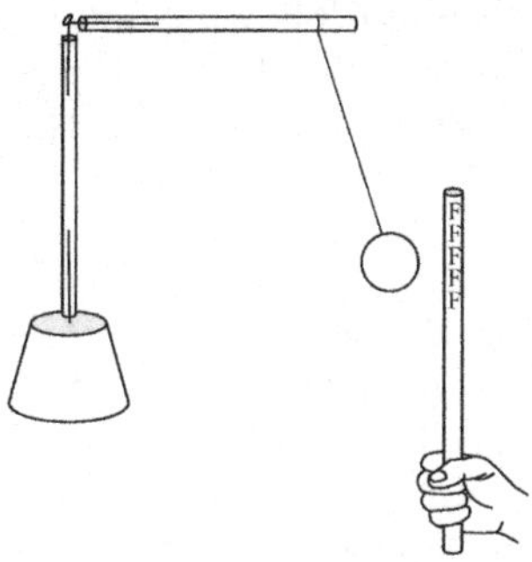

Figure 4.16: A rubbed plastic attracting the paper disk of a nearby electric pendulum.

This experiment is analogous to Experiment 2.1, showing an attraction between the rubbed straw and the pendulum. We now have a third criterion to say that a body is *electrically neutral*. The first criterion was presented in Experiment 2.1, that is, the body should not attract light bodies upward. The second criterion was that of Experiment 3.1: the body should not orient a metal versorium. The third criterion is the one illustrated here: it does not change the vertical angle of the thread of a nearby electric pendulum. A *charged body*, on the other hand, pulls light bodies upward, orients versoria, and changes the angle of the thread of an electric pendulum.

Experiment 4.6

We can make the paper disk follow the motion of the rubbed plastic by moving the plastic slowly to and fro in front of the pendulum. For the time being they should not be brought very close to one another, in order to prevent contact. In this case when the plastic moves toward the pendulum, the paper disk approaches the plastic, and when the plastic moves away from the pendulum, the paper disk comes back to the original vertical direction (Figure 4.17).

Experiment 4.7

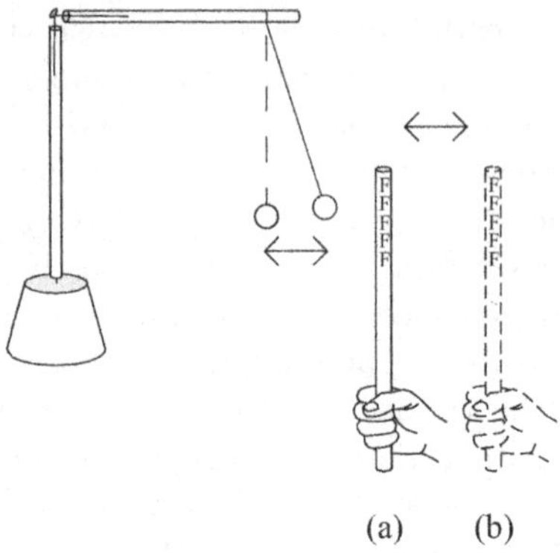

Figure 4.17: An electric pendulum following the motion of a rubbed piece of plastic. (a) When the straw is brought close to the paper disk, the paper moves toward it. (b) When the straw is moved away, the pendulum returns to the vertical direction.

We repeat Experiment 4.5. But now we bring the rubbed plastic closer to the paper disk, allowing them to come into contact. We observe that the pendulum is initially attracted and soon after is repelled by the rubbed plastic! Between the attraction and repulsion, something crucial happens: *contact* between the paper disk and the rubbed plastic. After this contact, the paper flies away from the rubbed straw whenever we try to bring them together (Figure 4.18).

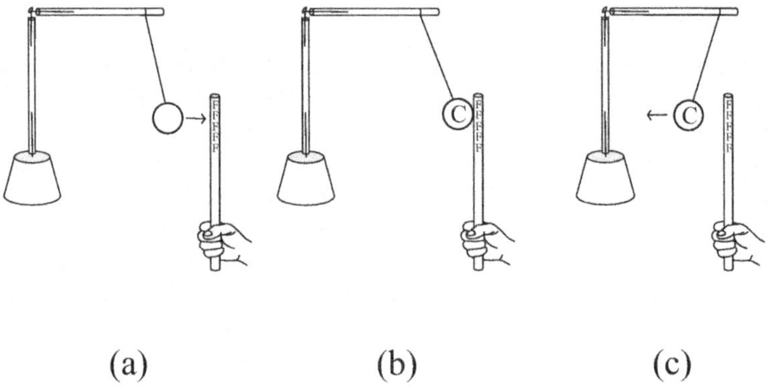

Figure 4.18: (a) The paper disk is initially attracted by the rubbed plastic, (b) touches it and afterward (c) is repelled by the straw.

Sometimes the paper disk is not immediately repelled by the rubbed plastic after contact, remaining in touch with it for a few seconds. In these cases we can observe the repulsion by tapping in the straw to release the disk. We can also move the straw up and down to release the paper, or blow it lightly. After release it is normally repelled by the rubbed plastic. In some cases there are necessary 2 or 3 attractions of the disk by the rubbed straw, always allowing their contact in each attraction, before we can observe their repulsion.

This repulsion after contact with a rubbed body may have occurred for some objects in Experiments 2.1, 2.3, and 2.4. But in these experiments the light objects, when no longer in contact with the rubbed plastic, fell to the ground due to the gravitational attraction of the Earth. And it is not easy to distinguish this attraction by the Earth with a possible repulsion exerted on them by the rubbed plastic which was placed above them. The advantage of the electric pendulum is that the silk thread balances the gravitational attraction exerted by the Earth on the paper disk. When the disk is repelled by the rubbed straw after contact, it will not fall to the ground, as it is being supported by the silk thread. It is then easy to see the repulsion between the rubbed plastic and the paper disk after their contact.

In any event, by performing Experiments 2.1, 2.3, and 2.4 once more, we can perceive the electrical repulsion after the contact acting together with gravitational attraction, provided we analyze all details of these experiments carefully.

Experiment 4.8

We now repeat Experiment 4.7. Initially the pendulum is attracted by the rubbed plastic, touches it, and begins to be repelled after release. We then remove the rubbed plastic and the disk returns to the vertical position. We now move a neutral wood skewer (or a sheet of paper, or our finger) near the paper disk. This motion of approach should be very slow, in order to prevent contact. What is observed is that the thread inclines toward the skewer, indicating that the paper disk is being attracted by the skewer (Figure 4.19).

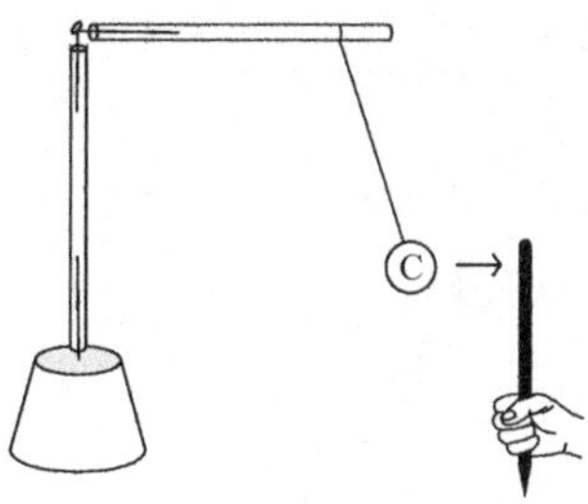

Figure 4.19: The electric pendulum which was being repelled by a rubbed straw after the two had been brought into contact, is now attracted by a neutral wood skewer.

As we saw in Section 3.5, this attraction indicates that the pendulum became electrically charged in Experiment 4.7. We were utilizing the letter F when a neutral plastic had become electrified by *friction*. We now observe that a paper disk becomes charged by mere *contact* with a rubbed piece of plastic. This charging process will be indicated by the letter C. This is the meaning of the letter appearing in Figures 4.19 and 4.18.

Definitions: We say that in Experiment 4.7 the paper disk of the pendulum *acquired an electrical charge due to contact with another charged body*, became *charged by contact*, or *electrified by contact*. The process is called *charging by contact*, *charge transference by contact*, or *electrification by contact*.

Instead of the word *contact*, people sometimes use a more generalized term. For instance, *electrification by communication* or *electrification by transference of charges*. There is a reason for this. Physical contact between the rubbed plastic and the paper disk is not always necessary in order to electrify the paper disk. Sometimes there is an electrical discharge (i.e., a spark) through the air when the rubbed plastic and paper disk are very close to one another. When this happens, there is a transference of charge between the rubbed plastic and the paper disk. In this case the paper disk, initially discharged, acquires an electrical charge and begins to be repelled by the rubbed plastic. In this book we will not deal with these phenomena of electrical discharges through the air.

4.5 Discharge by Grounding

Experiment 4.9

In order to repeat Experiment 4.8 with the same pendulum, we must first touch the paper disk with our finger. It is not necessary to hold the paper disk; a touch is enough. After this, when we bring another finger (or a sheet of paper, or a wood skewer) near the paper disk, we observe that the paper disk does not move, as the thread remains vertical. The paper disk is no longer attracted by the neutral skewer, finger, or sheet of paper, as was the case in Experiment 4.8. See Figure 4.20.

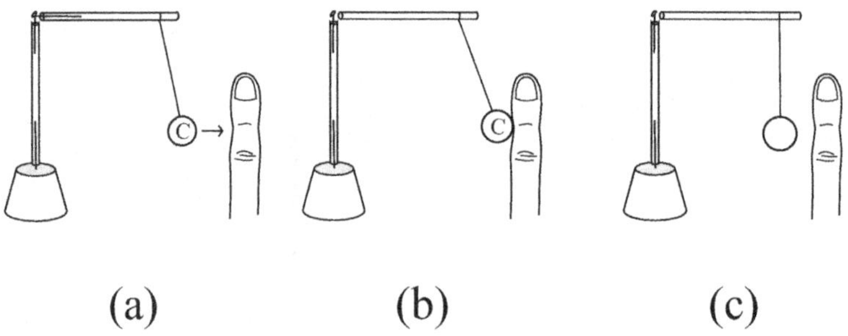

Figure 4.20: (a) A charged electrical pendulum is attracted by a finger. (b) If they touch one another, the pendulum is discharged. (c) After being discharged, the pendulum is no longer attracted by a nearby finger.

Definitions: We say that the paper disk *lost its electrical charge by contact* with the finger, or that it was *discharged by contact*, *electrically discharged* or, simply, *discharged*. The process is called *discharge by contact, by grounding, or by earthing*. It is also called *electrical grounding*, *electrical earthing*, *to ground*, or *to earth.* The origin of these names is that the charged body is being discharged by the human body, which is normally in electrical contact with the ground.

After discharge, when we bring a rubbed plastic near the pendulum, it is no longer repelled, as was the case at the end of Experiment 4.7. What is observed now is the same behaviour as at the beginning of Experiment 4.7. That is, the pendulum is initially attracted by the rubbed plastic, touches it, and only then is again repelled by it.

Whenever we touch the paper disk with our finger, we can repeat the whole procedure again. That is, the pendulum returns to its initial neutral condition.

Let us call the rubbed plastic of Experiments 4.7 and 4.9 body A. The paper disk will be called body B. And the hand which touches the disk will be called body C. It was in 1729 that Gray discovered that a body B, electrified by contact with a rubbed substance A, is discharged by contact with the hand C. One of his descriptions of the grounding effect appears in his work of 1731. He charged a long glass tube by friction. The tube was utilized to electrify another body by contact. We quote some of his words here (our emphasis in italics):[15]

> [...] for from several experiments it appears, that if any body [C] *touches* that [body B electrified by contact with a body A] which attracts, its attraction ceases till that body [C, which touched the electrified body B] be removed, and the other [body B, which was initially electrified and which was discharged by contact with body C] be again excited by the [rubbed glass] tube [A].

In 1733 Du Fay began to use this procedure systematically. He rubbed a glass tube and used it to electrify a wood ball:[16,17]

> Moreover, I was careful to touch the [electrified] ball with the hand after each experiment we had done with the tube [that is, after each electrification of the ball utilizing the rubbed glass tube], in order to remove [from the ball] all [electrical] virtue which it might have acquired by the approach of the [rubbed glass] tube; in effect, this [procedure] removed all his electricity, as mentioned by Mr. Gray; [...]

Experiment 4.10

We will now perform these experiments in a more detailed way in order to see all the effects. Initially we bring our finger near the uncharged paper disk of an electric pendulum. The disk does not move. We rub a straw and bring it near the disk. The disk is attracted, touches the straw and begins to be repelled by it, as in Figure 4.18. We remove the straw and the pendulum's thread returns to the vertical.

---

[15] [Grah, p. 35].

[16] [DF33d, p. 247].

[17] J'avois de plus le soin de toucher la boule avec la main après chaque station qu'on avoit faite avec le tube, afin de lui ôter toute la vertu qu'elle auroit pû avoir conservée par l'approche du tube; cela la dépouille en effet de toute son électricité, ainsi que l'a remarqué M. Gray; [...]

At this moment we move the finger slowly near the disk, but without touching. The disk is attracted by the finger, moving toward it.

If the disk touches the finger, the pendulum returns immediately to the vertical position. When we again bring the finger near the disk, it does not move, as it is no longer attracted by the finger. The pendulum has returned to its original situation.

We can rub the straw once more and begin this whole series of experiments over again.

4.6 Gray's Electric Pendulum

The oldest description we know of an electric pendulum has been given by Gray in 1720. He performed an experiment analogous to Experiment 4.5, but using a down feather instead of a paper disk:[18]

> A down feather being tied to the end of a fine thread of raw silk, and the other end to a small stick, which was fixed to a foot, that it might stand upright on the table: there was taken a piece of brown paper, which by the above-mentioned method [that is, the paper was initially warmed by the fire and then rubbed by passing it between his fingers] was made to be strongly electrical, which being held near the feather, it came to the paper, and I carried it with the same till it came near the perpendicular of the stick; then lifting up my hand till the paper was got beyond the feather, the thread was extended and stood upright in the air, as if it had been a piece of wire, though the feather was distant from the paper near an inch [2.54 cm].

We illustrate this experiment in Figure 4.21.

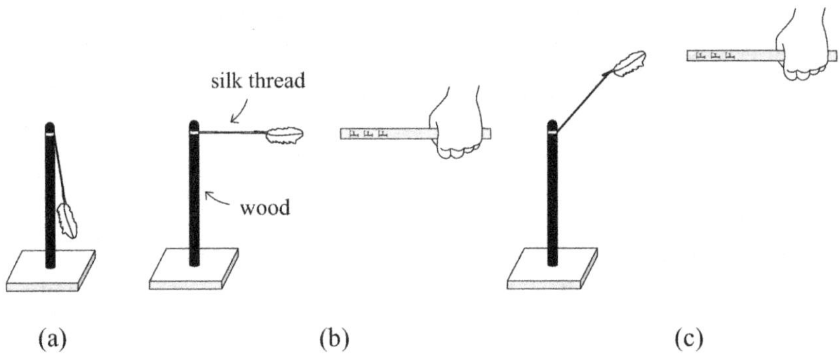

Figure 4.21: Gray's electric pendulum.

[18][Grab, p. 107].

4.7 The Du Fay Versorium

Another interesting way to observe electrical repulsion relies on a versorium created by Du Fay.[19] In his case it was a glass versorium with a hollow metal sphere at the end of one of its legs. Here we use an analogous plastic versorium, as described in Section 3.1, i.e., a plastic hat-shaped strip, with its two legs pointing slightly downward. There is a pin attached to the center of the plastic, with its tip downward and supported on the head of a nail stuck in a board. What characterizes Du Fay's versorium is that we wrap one of its ends with aluminum foil. There are two possibilities to balance the versorium if it begins to fall toward this side due to its extra weight. The first is to glue some plastic at the other end. The second possibility is to remove a small section of the leg where the aluminum foil will be attached before doing the experiment. The important point is that the versorium should be balanced horizontally, with one of its ends wrapped in aluminum foil (Figure 4.22).

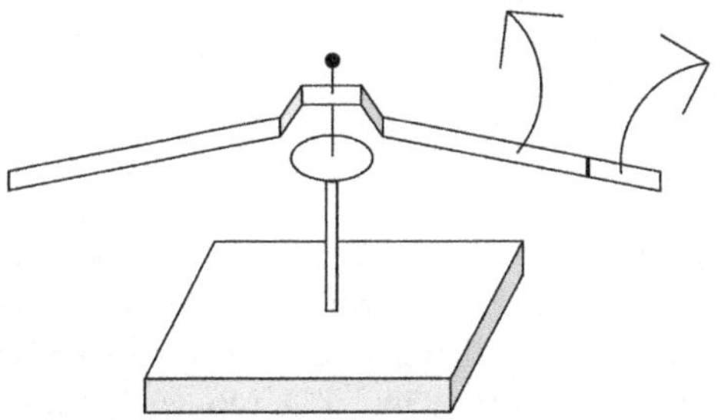

Figure 4.22: The Du Fay versorium is made of plastic, with the tip of one of its legs wrapped in aluminum foil.

It is a good idea to touch some metallic ground to insure one is neutral before doing the following experiments. In order to perform experiments with this versorium, it is important to be sure that it is initially neutral. This is the most delicate aspect, principally due to the plastic. In order to obtain charge neutrality we touch the aluminum foil with our finger to discharge it. We then place our finger close to several parts of the plastic versorium, without touching them. If the versorium remains at rest, without being oriented by our finger, we say that it is electrically neutral. When it reacts to the finger being brought close, we say it is charged. Sometimes we can inadvertently charge it by friction with our hand while building it or while we are wrapping the aluminum foil around one of its ends. If this happens, we can wait a few minutes so it discharges naturally. Alternatively, we can clean it with a wet paper napkin and wait for it to dry. Even after these procedures we must always check whether it is really neutral. From now on we will assume the versorium is neutral before starting the following experiments.

[19][DF33b, pp. 473-474].

Experiment 4.11

We rub a plastic straw and bring it slowly near the aluminum foil of Du Fay's versorium, not allowing them to come into contact. We observe that the versorium turns around its vertical axis, stopping with the aluminum foil pointing toward the rubbed plastic (Figure 4.23). When we move the straw, the versorium will follow it. This is analogous to Experiment 3.1.

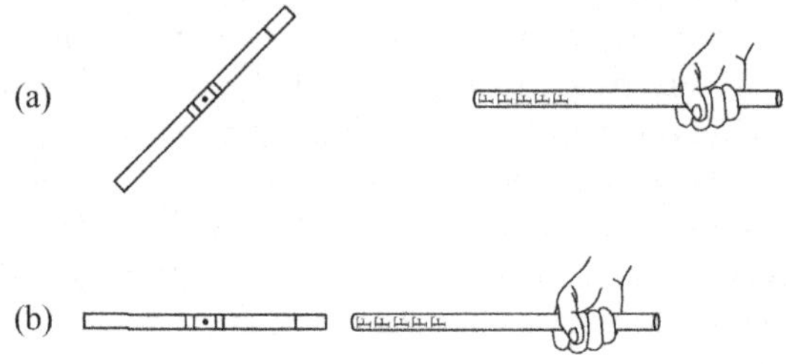

Figure 4.23: Aluminum foil paper of Du Fay's versorium being attracted by a rubbed plastic brought close to it.

Experiment 4.12

We repeat Experiment 4.11. But now we move the rubbed straw closer to the aluminum foil, allowing them to come into contact. We observe that it is first attracted by the straw, but is then repelled by it, pointing away from the straw (Figure 4.24). Between this attraction and repulsion something crucial happens: the *contact* between the aluminum foil and the rubbed plastic.

Sometimes the aluminum foil is not immediately repelled after contact, becoming attached to the straw. If this happens, it will be necessary to release the aluminum foil in order to see the following repulsion. This can be achieved by tapping on the straw, or by moving it up and down together with the aluminum foil. Normally they will detach from one another during this procedure. After release, the aluminum foil will begin to be repelled by the rubbed straw, moving away from it.

Experiment 4.13

We repeat Experiment 4.12. At the end of the experiment we remove the rubbed straw. We then slowly bring our finger (or a sheet of paper, or a wood skewer) near the aluminum foil, without allowing them to contact one another. We observe that the aluminum foil is attracted by the finger, pointing toward it and following its motion (Figure 4.25)!

As we saw in Section 3.5, this experiment indicates that the aluminum foil of Du Fay's versorium became charged when it touched the rubbed plastic.

A reminder is in order at this point. This experiment and several others described in this book may not work perfectly if it has rained recently or if the air is humid. Air humidity makes difficult for electrical charges to accumulate,

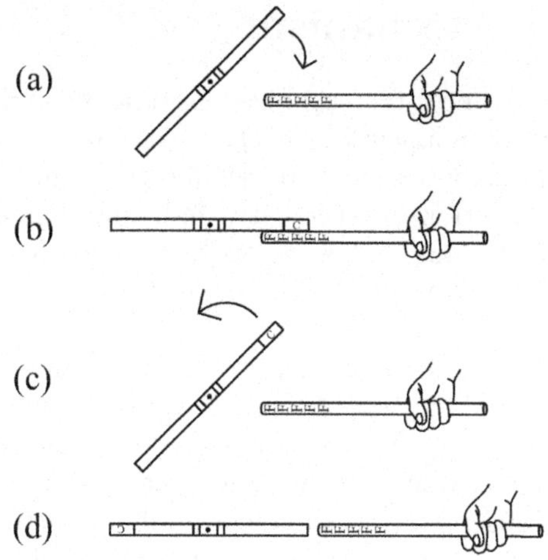

Figure 4.24: (a) The aluminum foil is initially attracted by the rubbed plastic. (b) The aluminum foil paper touches the rubbed portion of the straw. (c) After contact, the aluminum foil is repelled by the straw. (d) Final equilibrium orientation of the versorium.

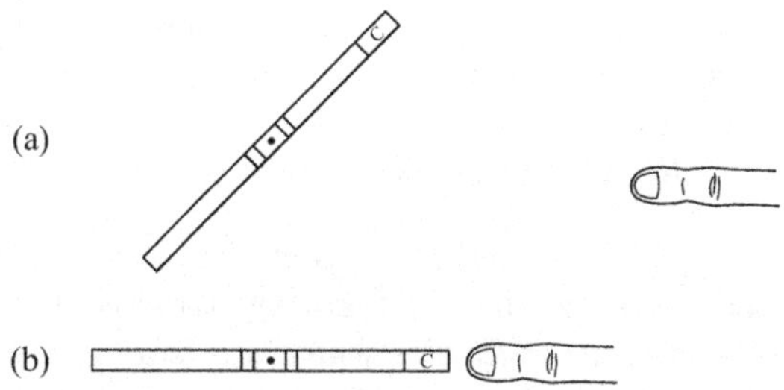

Figure 4.25: The aluminum foil which had previously touched a rubbed piece of plastic is now attracted by a finger moving near it.

and thus decreases the size of the effects which are to be observed. Electrostatic experiments normally work well in dry atmospheres.

Experiment 4.14

In order to repeat this whole series of experiments with the same Du Fay versorium, we must initially touch the aluminum foil with our finger. After this, when we again bring a finger or a sheet of paper near the aluminum foil we will observe that it no longer reacts and remains at rest. It is no longer attracted by the finger, as happened before in Experiment 4.13.

When we bring the rubbed straw near the aluminum foil again, we observe the same behaviour as before, namely, attraction, contact, and repulsion.

Whenever we touch the aluminum foil with our finger, we can start the whole procedure over again, since the versorium returns to its initial condition.

4.8 The *ACR* Mechanism

In 1733 Du Fay was the first to recognize the fundamental mechanism of attraction, contact (or communication of electricity by a close approach), and repulsion described in Experiment 4.10. Heilbron designated this simple rule of attraction, communication of electricity, and repulsion *ACR* (i.e., Attract, Communicate, Repel).[20] This regular behaviour was correctly regarded by Du Fay as a great discovery. After all, this principle explains a wide range of electric phenomena.

Du Fay had performed some earlier experiments and observed behaviour which enabled him to classify or distinguish different substances. When he rubbed a body and brought it near light substances, he observed that some of these substances underwent a stronger attraction than other substances of the same weight. He observed that the more easily attracted substances corresponded exactly to the substances which acquired a smaller charge by friction. An example of this behaviour was seen in Sections 2.4 and 2.7. The small píeces of plastic or silk, for instance, are much less attracted by a rubbed plastic than small pieces of metal or paper. On the other hand, we can electrify plastic and silk more easily by friction than we can paper or metal.

In Du Fay's time bodies which had the property of attracting light substances when rubbed were commonly called *electric*. An electric body was considered good or bad according to the greater or smaller force with which it attracted light bodies after being rubbed. After this introduction, we can quote Du Fay's own words describing the *ACR* principle:[21,22]

> Then, having thought about the fact that the bodies which are less electric by themselves were more vigorously attracted [by the electrified bodies] than the other [bodies which are more easily electrified by friction], I imagined that perhaps the electric bodies would attract all bodies which are not [electric, that is, he imagined that the bodies which are electrified by friction would attract all bodies which we cannot electrify by friction], and would repel all bodies which become [electrified] by its approach [that is, by the approach or by the contact with the electrified body], and by communication of the [electrical] virtue.
>
> [...]

Du Fay continues and describes experiments he performed that were analogous to experiments of Guericke, Gray, and Hauksbee. (See Experiment 4.4). That is, he could float a piece of gold leaf which had been released over a rubbed

---

[20] [Hei99, pp. 5 and 255-258].

[21] [DF33b, p. 458].

[22] "Enfin ayant refléchi sur ce que les corps les moins électriques par eux-mêmes étoient plus vivement attirés que les autres, j'ai imaginé que le corps électrique attiroit peut-être tous ceux que ne le sont point, & repoussoit tous ceux que le sont devenus par son approche, & par la communication de sa vertu.

[...]

glass tube in air. The leaf was first attracted by the tube, touched it, and afterwards was repelled by it, becoming floating above it. He went on to write: (Our emphasis in italics)[23,24]

> The explanation of all these facts is very simple, by supposing the *principle* which I just quoted; for, in the first experiment, when the [gold] leaf is released over the [rubbed glass] tube, it attracts strongly this leaf which is not at all electric, but after it has touched the tube, or that it had simply approached the tube, it is repelled by it, and is always kept away from it, until the small electric vortex it had acquired has been exhausted, or at least greatly reduced; then, being no longer repelled, it falls once more over the tube, where it acquires a new vortex [a new electric charge] and, consequently, new forces to avoid the tube, [and this process] continues while the tube maintains its [electrical] virtue.

Here is another description of this *principle*:[25]

> On making the experiment related by Otho de Guerik, in his collection of experiments *de Spatio Vacuo* [1672], which consists in making a ball of sulphur render'd electrical, to repel a down-feather, I perceived that the same effects were produced not only by the [rubbed] tube, but by all electric bodies whatsoever; and I discovered a very simple principle, which accounts for a great part of the irregularities, and if I may use the term, of the caprices that seem to accompany most of the experiments on electricity. This principle is that electrick bodies attract all those that are not so, and repel them as soon as they are become electrick, by the vicinity or contact of the electrick body. Thus leaf-gold is first attracted by the [rubbed glass] tube; and acquires an electricity by approaching it; and of consequence is immediately repell'd by it. Nor is it re-attracted, while it retains its electrick quality. But if, while it is thus sustain'd in the air, it chance to light on some other body, it straightways loses its electricity; and consequently is re-attracted by the tube, which, after having given it a new electricity, repels it a second time; which continues as long as the tube keeps its electricity. Upon applying this principle to the various experiments of electricity, one will be surprized at the number of obscure and puzzling facts it clears up.

---

[23] [DF33b, pp. 459-460].

[24] L'explication de tous ces faits est bien simple, en supposant le principe que je viens d'avancer; car, dans le premiére expérience, losqu'on laisse tomber la feuille sur le tube, il attire vivement cette feuille qui n'est nullement électrique, mais dès qu'elle a touché le tube, ou qu'elle l'a seulement approché, elle est renduë électrique elle-même, & par conséquent elle en est repoussée, & s'en tient toûjours éloignée, jusqu'à ce que le petit tourbillon électrique qu'elle avoit contracté soit dissipé, ou du moins considérablement diminué; nétant plus repoussée alors, elle retombe sur le tube où elle reprend un nouveau tourbillon, & par conséquent de nouvelles forces pour l'éviter, ce qui continuëra tant que le tube conservera sa vertu."

[25] [DF, pp. 262-263].

Experiment 4.15

We can make the electric pendulum oscillate by performing a curious experiment. We place the left hand with four fingers close to the paper disk, without touching it. The hand will always remain in this position during all the experiment. We rub a plastic straw and bring it slowly near the paper disk. After the disk touches it, the straw should remain at rest.

We observe that the pendulum is initially attracted by the rubbed plastic, touches it, is repelled by it, moves toward the hand, touches it, and is again attracted by the straw. The whole process repeats for a few quick oscillations of the paper, which alternately touches the rubbed plastic and the hand on the other side. We can increase the number of these oscillations by turning the straw around its axis during the experiment, or by moving the straw vertically along its length during the oscillations (Figure 4.26).

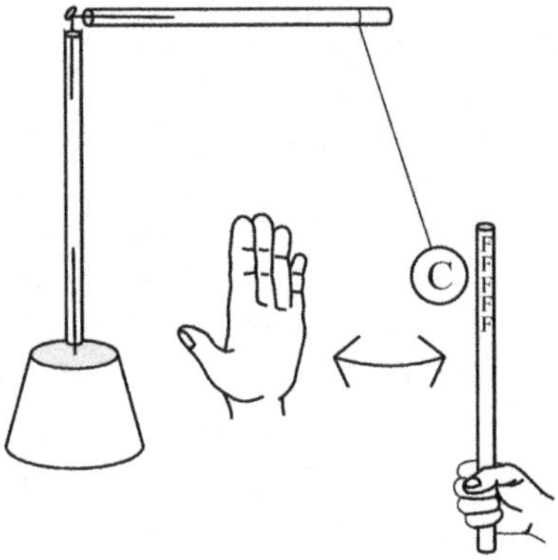

Figure 4.26: Pendulum oscillating between a rubbed plastic and hand, and touching each of them.

It is possible to describe what is happening in this experiment by referring to the *ACR* mechanism. This experiment is analogous to Gray's floating feather, which oscillates in air between the rubbed glass tube and a solid object (like a wall or a chair). See Section 4.1.

Experiment 4.16

We now place two electric pendulums like the one used in Experiment 4.7 side by side. When both of them are neutral, the two threads remain vertical. We can make this happen by touching both paper disks with our hand.

We rub a plastic straw and bring it near each paper disk, allowing them to touch the straw and be repelled by it. We now remove the straw. After this, we move the pendulums which had been charged by contact closer to one another. We observe that they repel one another. Both threads incline relative to the vertical, moving away from one another (Figure 4.27).

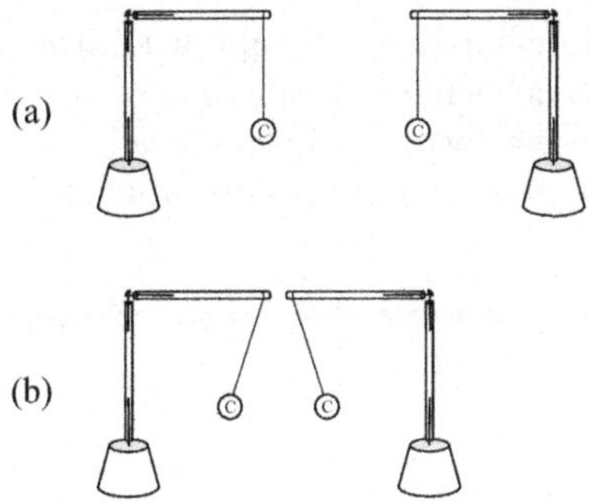

Figure 4.27: (a) Two charged pendulums hang vertically when they are far away from one another. (b) Two charged pendulums repel one another when close together.

Sometimes it is necessary to use very thin silk threads in order to see this lateral separation. When the threads are very dense and heavy, their weight decreases the angle of separation. Moreover, shorter threads create greater angles of separation between them than longer threads, for the same final distance between the paper disks. This means that it helps to work with short threads as this makes the repulsion more visible.

In this experiment we are seeing the repulsion between two pendulums which had been charged by contact with a single electrified body.

This experiment also illustrates the action and reaction between two electrified bodies, a subject discussed previously in Section 3.5.

Experiment 4.17

An analogous experiment can be made substituting two small crumpled balls of aluminum foil for the paper disks. Each ball can be made of a square or a circle with a 2 to 3 cm side or diameter. When they have been crumpled, they are tied to the ends of two silk threads of the same length hanging like pendulums. We charge both balls by contact with a rubbed plastic, which is then moved far away. After this procedure, they repel one another when the pendulums are placed close together. The shorter the threads, the greater will be the angle of separation for a constant distance between the upper ends of the threads.

4.9 Gray's Pendulous Thread

Apart from the electric pendulum, it is also interesting to make another instrument which is called "pendulous thread." It was created by Stephen Gray in 1729, as a means of detecting whether a body is charged.[26]

[26] [Grad], [Graf], and [Grai].

It is simply a cotton or linen thread supported from above by a wood stick (Figure 4.28). The electric pendulum was made with a silk or nylon thread. Here it is important to use a cotton or linen thread. We can hold the stick with our hand or attach it to another appropriate support.

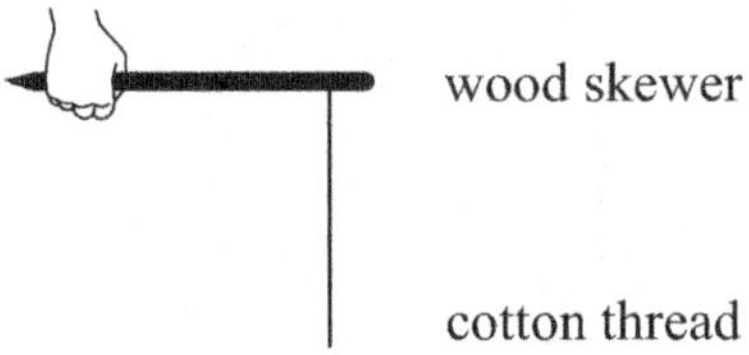

Figure 4.28: Gray's pendulous thread.

Experiment 4.18

We move a neutral piece of plastic near a pendulous thread. Nothing happens, as it remains vertical. We now bring a rubbed piece of plastic near a pendulous thread. The thread inclines toward it (Figure 4.29).

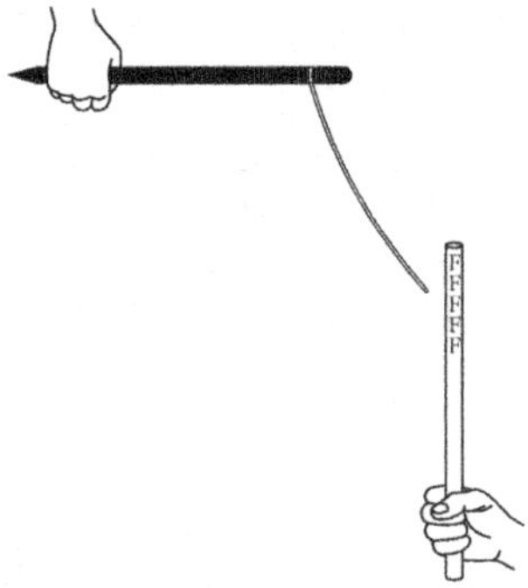

Figure 4.29: Attraction of a pendulous thread by a rubbed plastic.

This is the main utility of the pendulous thread. That is, it tells us whether or a nearby object is charged. Gray's pendulous thread was simply a vertical thread supported by its upper end, with its lower end free to move in any direction. The thread was made of cotton or linen, without any feather or other body at its lower end. In order to test whether a body was charged, he simply brought the thread close to it. When the thread was attracted by the body, inclining toward it, this meant that the object was electrically charged. The normal procedure for determining whether or not a body was charged, was to observe if it attracted nearby light substances, as in Experiment 2.1. With his instrument Gray had found a new method, the inclination of his thread. He explicitly mentioned that this new method allowed greater sensitivity than the older method:[27]

[27][Graf, p. 289].

> The manner of observing these attractions is best performed by holding the attracting body in one hand, and a fine white thread tied to the end of a stick, in the other [hand]; by this means far less degrees of attraction will be perceived, than by making use of [pieces of] leaf-brass.

Experiment 4.19

We now allow the pendulous thread to touch the rubbed plastic. We observe that it remains stuck to it (Figure 4.30).

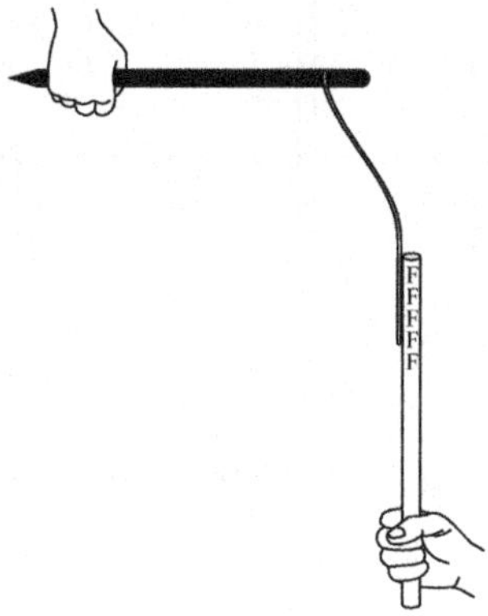

Figure 4.30: Gray's pendulous thread remains stuck to a rubbed plastic after contact.

This differs from what happened in Experiment 4.7. In this earlier experiment the electric pendulum was repelled after contact with the rubbed plastic. This means that the electric pendulum and the pendulous thread are different instruments, which present different behaviours in analogous situations. The pendulous thread is not simply an electric pendulum without the paper disk.

4.10 Mapping the Electric Force

We can adapt an electric pendulum to map the electric force, in analogy with what was done with the versorium in Section 3.4. To do this, we have to make an electric force indicator. This is essentially an electric pendulum in which we replace the paper disk with a small arrow made of paper, aluminum foil, or thin cardboard. It should point horizontally and be suspended at its center by a silk or nylon thread. It can be 2 to 5 cm long, with a vertical shaft width from 0.2 to 0.5 cm, and the maximum width of the arrow tip from 0.5 to 0.7 cm. These are only approximate measures and are not so critical.

A practical way to make and attach the arrow is with a plastic straw.[28] The thin cardboard arrow can be initially 4 to 6 cm long, with a shaft width of 0.2

[28] [FM91].

to 0.5 cm, and the maximum dimension of the arrow tip from 0.5 to 0.7 cm, for instance. We cut a plastic straw 3 to 5 cm long. We place glue over one side of the arrow, and glue the lower section of the silk thread and the straw on this side. We fold the arrow's shaft around the straw, attaching it them together. It should lie horizontally when hanging freely. If this does not happen, we can cut a section off the straw in order to have the arrow in the correct position. Several electric force indicators like this one should be built (Figure 4.31).

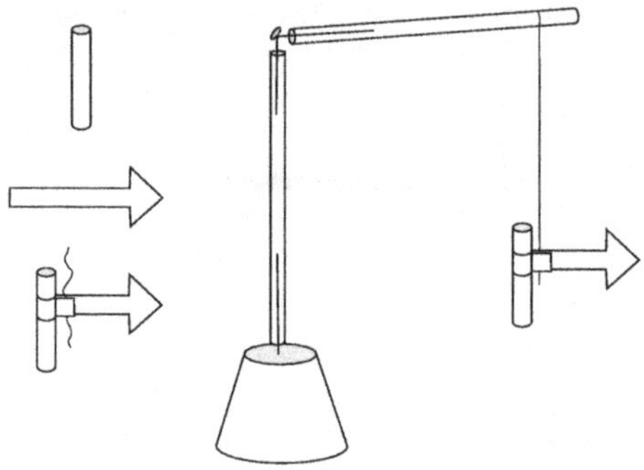

Figure 4.31: An electric force indicator.

Experiment 4.20

We now repeat the procedure of Experiment 4.5. That is, we rub a plastic straw which is set vertically in an appropriate support. We then bring it slowly near the electric force indicator, initially preventing the arrow from touching the rubbed plastic. We observe that the pendulum is attracted by the rubbed plastic, with the thread inclined toward it. Moreover, the arrow tip points toward the rubbed straw, regardless of its position relative to the rubbed plastic. This shows that the electric force exerted by the rubbed plastic points toward it, as we saw in Experiment 3.4.

Experiment 4.21

We now repeat Experiment 4.20, this time allowing the arrow and rubbed plastic to come into contact. After contact, the pendulum is repelled by the straw, with the thread inclined away from it. Moreover, the arrow tip points radially away from the rubbed plastic (Figure 4.32).

Experiment 4.22

We now use several electric force indicators around a rubbed plastic. The initial procedure is like that of Experiment 4.20, namely, preventing the contacts between the arrows and the straw. We observe that all of them point toward the rubbed plastic (Figure 4.33 (a)). We now allow the rubbed straw and arrows to come into contact. We observe that after contact all of them point radially away from the straw (Figure 4.33 (b)).

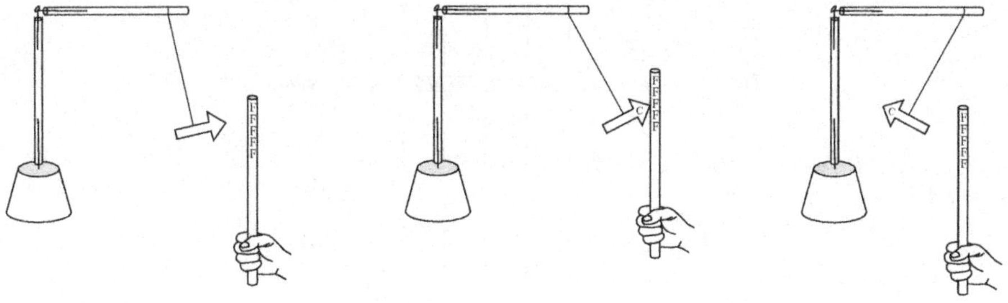

(a) (b) (c)

Figure 4.32: (a) Initially the arrow points toward the rubbed plastic, as it is attracted by it. (b) It touches the straw. (c) After contact, the arrow is repelled by the straw, with the arrow tip pointing away from the straw.

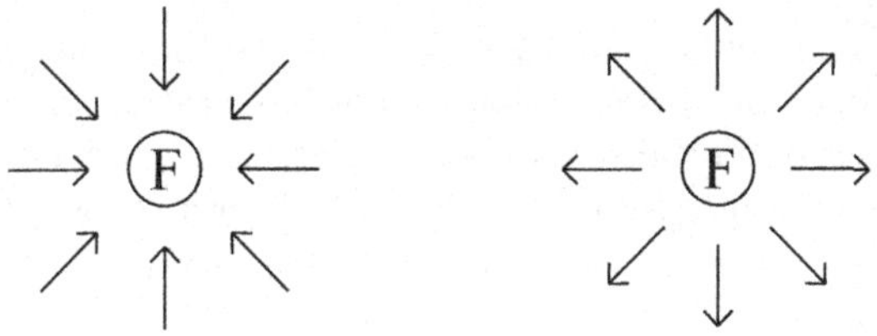

(a) (b)

Figure 4.33: (a) Before contact the arrows point toward the rubbed straw, being attracted by it. (b) After contact they point away from the plastic, being repelled by it.

The main difference between Experiments 4.20, 4.21, and 4.22, on the one hand, and Experiment 3.4 on the other, is that now the arrows indicate not only the direction of the force (in this case a radial direction), but also whether the force is attractive (arrows pointing toward the rubbed straw) or repulsive (arrows pointing away from the rubbed straw).

Experiment 4.23

The same experiment can be repeated with the rubbed straw in a horizontal position. Before contact the arrows point toward it, after contact away from it, as in Figure 4.34.

Experiment 4.24

We now rub two plastic straws along their lengths with the same material, like a sheet of paper. These straws are placed side by side vertically on appropriate supports. The electric force indicator is moved near the straws, without

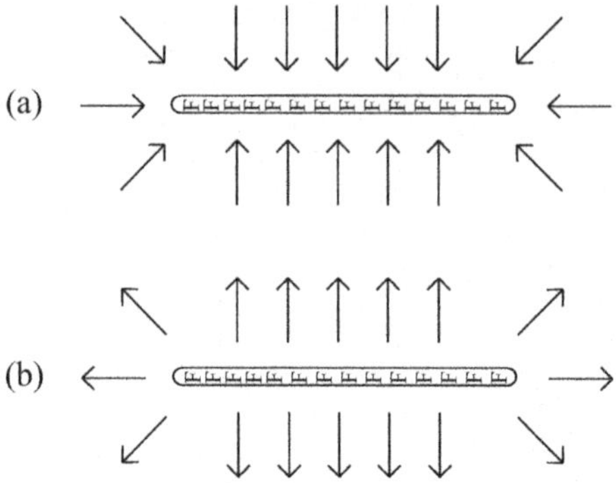

Figure 4.34: (a) The arrows which did not touch the rubbed straw are attracted by it. (b) After contact they are repelled by it.

allowing them to contact one another. We observe that the arrow is attracted by both straws, inclining toward them. The orientation of several arrows in different locations around the straws is presented in Figure 4.35 (a). The orientation of each arrow results from the influence of both straws. It is like a vector addition of the forces or torques exerted by each straw. This is analogous to Experiment 3.5.

We repeat this experiment but now allowing the contact between the straws and the rubbed plastics. After contact the arrows are repelled by the straws, as indicated in Figure 4.35 (b).

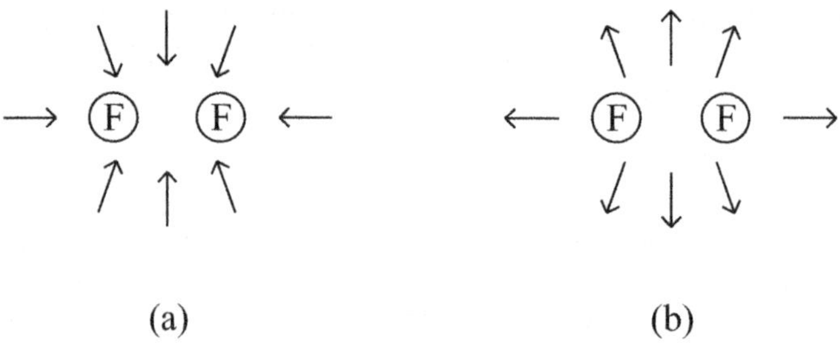

Figure 4.35: Electric force indicators being attracted, (a), and repelled, (b), by two rubbed plastics.

These experiments indicate the vectorial nature of electric forces, whether they are attractive or repulsive. The advantage of these mappings with arrows, in comparison with the mapping with versoria, is that the arrows indicate not only the direction of the forces, but also if they are attractive or repulsive.

Experiment 4.25

The same effect can be obtained with several Du Fay versoria, instead of simple metal versoria. We rub a plastic and place the rubbed part at the same level as the plane of the versoria. We move the rubbed plastic near the versoria, preventing their touch. The versoria rotate around their axes. After reaching equilibrium and stopping, the aluminum foils of the versoria point toward the rubbed plastic (Figure 4.36 (a)).

We now allow contact between the rubbed plastic and the aluminum foils, until they are repelled by the plastic. The versoria rotate and, in the new equilibrium positions, the aluminum foils point away from the rubbed straw (Figure 4.36 (b)).

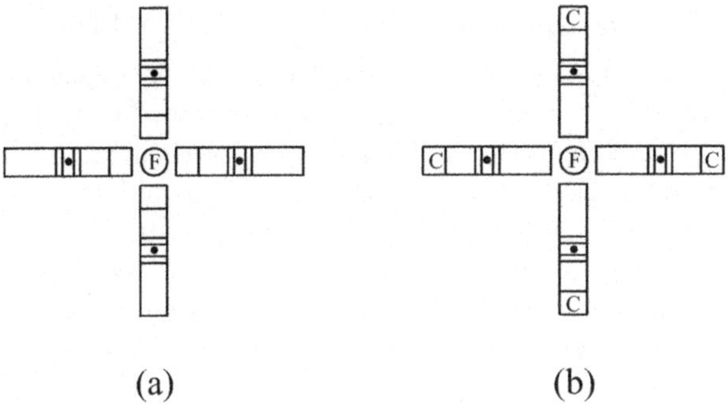

Figure 4.36: (a) The aluminum foil of each Du Fay versorium is attracted by a rubbed piece of plastic. (b) After contact it is repelled by the charged straw.

With Du Fay's versoria we can also obtain mappings analogous to those of Experiments 4.23 and 4.24.

4.11 Hauksbee and the Mapping of Electric Forces

Probably the first person to map electric forces was Hauksbee in 1706. He utilized the electrical machine which he invented, described in Section 4.2 (Figure 4.8). He replaced the glass globe with a cylindrical glass tube, supported horizontally or vertically, so that it could rotate at high speed on its axis. While spinning, it was rubbed with the hands or with a sheet of paper. Here is the description of his experiment:[29]

> *A continuation of the experiments on the attrition of glass.*
>
> I procur'd a glass nearly cylindrical, of the length and diameter about seven inches each [18 cm], whose motion [rotation around its axis of symmetry] was given by a machine of a new contrivance; its axis lying parallel to the horizon, which in like experiments heretofore made, was diametrically opposite to it. [...]

[29][Haua, pp. 2332-2335].

Now what farther I have to add, occurr'd from observing always that light bodies, would seemingly be equally attracted, or gravitate; so that I contrived a semi-circle of wire, which I could fasten at a constant distance [from the axis of the cylinder], environing the upper surface of the glass at 4 or 5 inches [10 or 13 cm] from it. This wire had twisted round it some pack-thread, whereby I could with ease hang the [linen, cotton, or woolen] threads at pretty nearly equal distances; the lower ends of which reaching within less than an inch [2.54 cm] of the glass, when held approaching the center of it, but appear'd, when at liberty, as Figure the 1st [of Figure 4.37].[30]

And when the cylinder was pretty swiftly turn'd about, those threads would appear by the agitated air, as in Fig. the 2d. But when on the lower part of the [spinning] glass was applyed my hand [in order to charge the glass by friction], the threads would then represent a form like Fig. the 3d. And from all parts seem to gravitate, or were attracted in a direct line to the center of the moving body [that is, towards the axis of the cylinder], suffering no inconvenience or disorder of posture by the wind occasioned by the rapidity of the motion; and I could by shifting the [position of] attrition, draw them in a line towards either end of the cylinder; yet still pointing to the axis of it. And if the [semi-circular] wire with the threads be revers'd, as I have tryed since, that is, encompassing the under part of the cylinder, as before the upper, it answer'd exactly the same as the other; the threads all pointing to the axis of it: See Fig. the 4th [Figure 4.37]. I have likewise given a motion to the same glass in a perpendicular posture, by which means I had the opportunity of placing a hoop-wire horizontally, with threads as before, and left only one small part expos'd for the touch of my fingers between them; yet the threads upon the motion and attrition given the cylinder, elevated themselves from their hanging posture, making all round an horozontal plain, directing their loose ends to the axis as in the other. Now how far this experiment may serve to explain the nature of electricity, magnetism, or gravatation of bodies, is beyond my sphere to determine; but with all humility submit it to those learned Gentlemen of this honourable Society, who have already treated on those subjects.

[30] [Haua] and [RR57, p. 568].

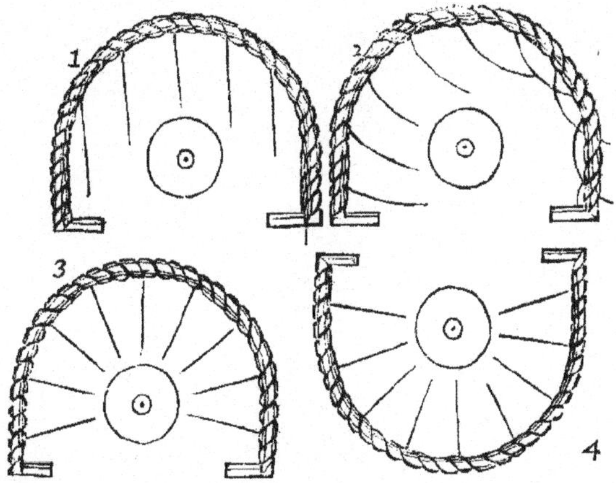

Figure 4.37: Hauksbee's electric force mapping. End view of the glass cylinder showing the positions of the threads when the cylinder is: (1) motionless and unelectrified; (2) rotating and unelectrified, the air currents around it dragging the threads all in the same direction; (3) and (4), rotating and electrified, in which case, despite the continued presence of the air currents, each thread straightens out and points to the axis of the cylinder.

Chapter 5

Positive and Negative Charges

5.1 Is There Only One Kind of Charge?

We now perform some experiments analogous to Experiment 4.7. We utilize the following instruments: Two electric pendulums, labeled I and II, and a metal versorium. Moreover, we will also use the following materials: Two plastic straws, two acrylic rulers, two glass cups, two silk stockings, and two pieces of cloth made of acrylic threads. These materials and their shapes as represented in the following experiments are described in Figure 5.1. Instead of two silk stockings we can also use two tissues made of synthetic polyamide.

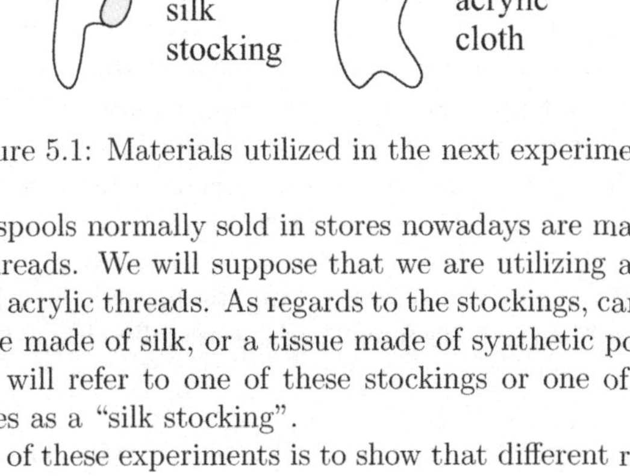

Figure 5.1: Materials utilized in the next experiments.

The "wool" spools normally sold in stores nowadays are made of 100% synthetic acrylic threads. We will suppose that we are utilizing a cloth or blouse made with these acrylic threads. As regards to the stockings, care must be taken to use only those made of silk, or a tissue made of synthetic polyamide. In the next figures we will refer to one of these stockings or one of these synthetic polyamide tissues as a "silk stocking".

The purpose of these experiments is to show that different rubbed materials exhibit different electrical charges. Before beginning each experiment we must touch the versorium and the paper disks of the electric pendulums with our finger in order to discharge them. This should be repeated before moving each of the rubbed objects near the pendulums. The versorium will be utilized to test whether the bodies are neutral or charged. One straw, one cup, one stocking and one ruler should be neutral, not affecting the versorium, and they will not be rubbed during the experiment.

In order for glass to be electrified, it usually needs to be dry. Moreover, the glass should be warmed before it is rubbed; otherwise it will discharge through our hand. Handling leaves sweat on the glass, and this should be avoided. It can be warmed by fire or in a microwave before each friction process. If one glass cannot be charged, you may have to try other glass blends, or other types of glass, until you find an adequate chargeable glass.

The acrylic cloths will be utilized to rub all these materials. We can hold the plastic straw (or the acrylic ruler, or the silk stocking) with the acrylic cloth and pull quickly the straw. We can also rub the cloth quickly over the surface of the glass.

When we rub an object with the silk stocking, this will be represented by the letter S on the body. When this object is rubbed with an acrylic cloth, this will be represented by the letter A. We will use two letters on paper disks which have first touched the rubbed bodies and subsequently were repelled by them. The first is to represent the substance of the object touched by paper disk. The second letter represents the material with which the object was rubbed. The material making up the objects will be represented by the letters P, G, A, and S. They indicate, respectively, plastic, glass, acrylic, and silk. For instance, the letters PA on a paper disk indicate that it touched a piece of Plastic which had been rubbed with Acrylic cloth, and the paper disk was then repelled by the rubbed plastic.

We first touch the versorium and the two paper disks of the pendulums with our finger. We bring the unrubbed straw (or glass, stocking, ruler, or cloth) near the versorium, observing that all these bodies are electrically neutral, as they should not orient the versorium. When any one of these unrubbed materials orients the versorium, it should be replaced by another unrubbed material which does not orient the versorium and is thus really electrically neutral. During the experiment we will rub a straw, a glass, a stocking, and a ruler. The following experiments only work when these objects are successfully charged by friction. In order to be sure that this charging mechanism has worked, we will move each of these objects near the versorium, before moving them near the pendulums. We should continue with the experiment only when the versorium orients toward these bodies. This precaution is especially relevant in the case of glass. As mentioned before, it is not always easy to keep a rubbed glass electrified. Due to the contact with our hand it can be easily discharged. From now on we will suppose that all rubbed objects have been successfully charged.

Experiment 5.1

A plastic straw which has been rubbed with an acrylic cloth moves near the first neutral pendulum. The pendulum is attracted, touches the straw, and is then repelled by it (Figure 5.2 (a)). We remove the plastic and the pendulum returns to the vertical. We rub a silk stocking with another acrylic cloth. We bring this rubbed stocking near the second neutral pendulum. It is attracted by the stocking, touches it, and is then repelled by it (Figure 5.2 (b)). We remove the stocking and the pendulum returns to the vertical.

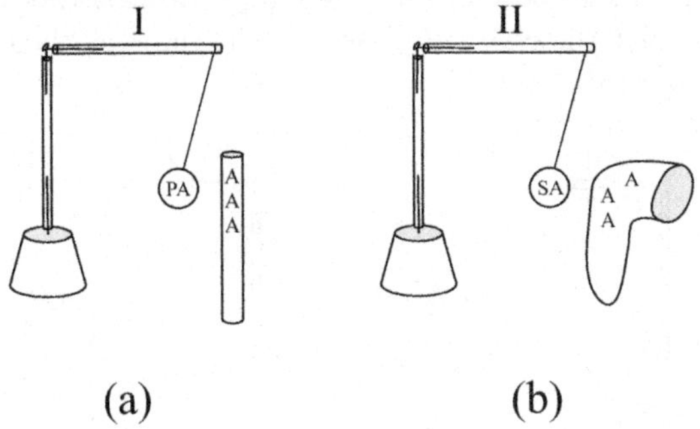

Figure 5.2: After contact the pendulums are repelled by objects that have touched them.

We now slowly move the rubbed silk near the first pendulum, not allowing them to contact one another. We observe a strong attraction between them (Figure 5.3 (a))! When we slowly bring the rubbed straw near the second pendulum, again not allowing them to contact one another, there is another strong attraction (Figure 5.3 (b)).

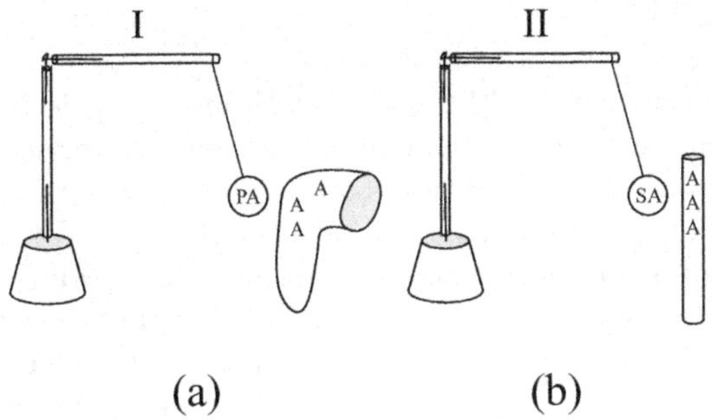

Figure 5.3: (a) The first pendulum, charged by contact with a plastic rubbed with acrylic cloth, is attracted by a silk stocking rubbed with acrylic cloth. (b) The second pendulum, charged by contact with a stocking rubbed with acrylic cloth, is attracted by a plastic straw rubbed with acrylic cloth.

Experiment 5.2

The glass cup is warmed and rubbed with an acrylic cloth. The rubbed portion of the glass is moved slowly near the first pendulum, which had been charged by contact with the straw in Experiment 5.1, and the glass and the paper disk of the pendulum are not allowed to come into contact. We observe a strong attraction between the rubbed glass and the charged pendulum (Figure 5.4 (a)). On the other hand, when the rubbed portion of the glass is slowly brought near the second pendulum, which had been charged by contact with the silk stocking in Experiment 5.1 (no contact between the glass and the paper disk of the pendulum), we observe that they repel one another (Figure 5.4 (b)). Thus we conclude that the charged glass acts the same as the charged stocking and not like the charged straw.

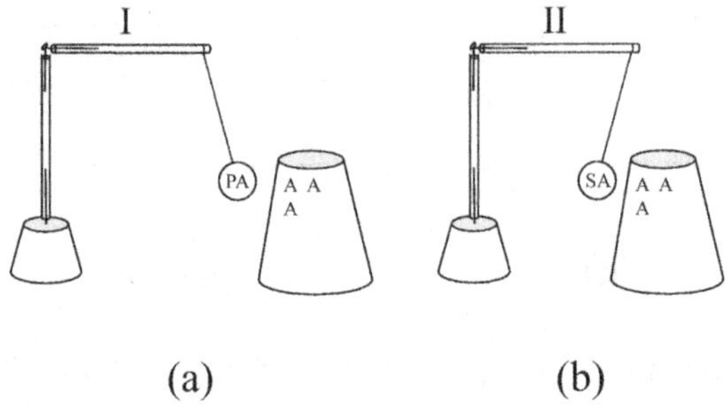

Figure 5.4: (a) The first pendulum, charged by contact with a plastic rubbed with acrylic cloth, is attracted by a glass rubbed with acrylic cloth. (b) The second pendulum, charged by contact with a silk stocking rubbed with acrylic cloth, is repelled by a glass cup rubbed with acrylic cloth.

Experiment 5.3

We rub the acrylic ruler in the acrylic cloth and the ruler is slowly brought near the first pendulum, which had been charged by contact with the straw rubbed with acrylic cloth, preventing contact between the ruler and the pendulum. We observe that they repel one another (Figure 5.5 (a)). On the other hand, when the rubbed ruler is slowly brought near the second pendulum, which had been charged by contact with the silk stocking rubbed with acrylic cloth, they are strongly attracted (Figure 5.5 (b)). Thus we conclude that the charged ruler acts the same as the charged straw and not like the charged stocking or the charged glass.

The order of this experiment can be reversed and the result is the same. For instance, we discharge the pendulums, rub the acrylic ruler in an acrylic cloth and this ruler charges the first pendulum by contact. The glass cup is warmed, then rubbed with an acrylic cloth and it charges the second pendulum by contact. When the ruler (or the straw) rubbed with acrylic cloth is slowly moved near the first pendulum, without contact, a repulsion results. When this ruler (or straw) is slowly brought near the second pendulum, once more without contact, an attraction results. On the other hand, when we slowly bring the glass (or silk) rubbed with acrylic cloth near the first charged pendulum, without

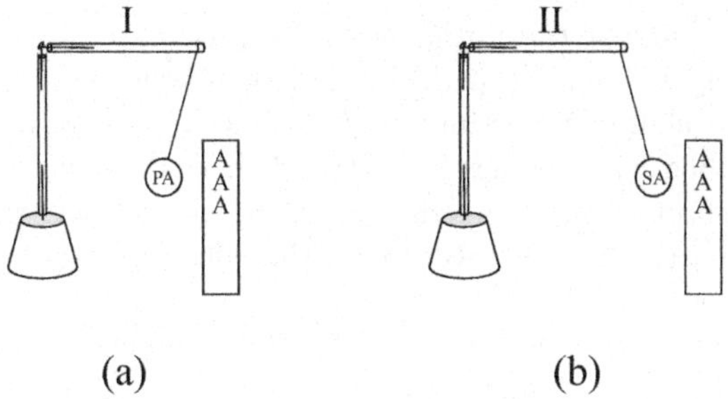

Figure 5.5: (a) The first pendulum, charged by contact with a plastic rubbed with acrylic cloth, is repelled by an acrylic ruler rubbed with an acrylic cloth. (b) On the other hand, the second pendulum, charged by contact with a silk stocking rubbed with acrylic cloth, is attracted by an acrylic ruler rubbed with an acrylic cloth.

contact, a strong attraction results. Slowly bringing the glass (or silk) rubbed with acrylic cloth near the second charged pendulum, without contact, yields a repulsion.

These experiments can be repeated with other materials yielding similar results. There are always attractions or repulsions between bodies charged by friction and pendulums charged by contact. And the charged bodies can be divided into two separate groups. In our example, the first group is made up of the plastic straw rubbed with acrylic cloth, the acrylic ruler rubbed with acrylic cloth, the electric pendulum charged by contact with a straw rubbed with acrylic cloth or with a ruler rubbed with acrylic cloth. The second group consists of the glass rubbed with acrylic cloth, the silk stocking rubbed with acrylic cloth, the electric pendulum charged by contact with the glass rubbed with acrylic cloth or with the silk stocking rubbed with acrylic cloth.

Experimental observations: What is observed is the following: objects in the first group repel one another; objects in the second group repel one another, and objects in different groups attract one another.

Definitions: Objects in the first group are said to be *negatively charged*, *negative*, or we say that they have acquired a *negative charge*. Objects in the second group are said to be *positively charged*, *positive*, or we say that they have acquired a *positive charge*. In the following Figures we represent this convention with the symbols "−" and "+", respectively.

There are repulsions between bodies having charges of the same sign in Figures 5.2 (a) and (b), 5.4 (b), and 5.5 (a). There are attractions between bodies charged with opposite charges in Figures 5.3 (a) and (b), 5.4 (a), and 5.5 (b).

Experiment 5.4

The attractions described in Experiment 5.1 are different from the attractions occurring between a charged pendulum and a neutral body. In order to verify this fact, we repeat the initial portion of the experiment, charging pendulum I negatively by contact with a negative plastic straw (rubbed with an acrylic cloth), and charging pendulum II positively by contact with a positive silk stocking (rubbed with an acrylic cloth). When we slowly bring the rubbed straw near pendulum II, without allowing them to contact, we observe an attraction much greater than the attraction that occurs between this pendulum and a neutral straw.

The force intensity can be measured by three quantities, namely, (a) minimum distance, (b) angle of inclination for a fixed distance between the straw and the vertical passing through the point of support for the thread of the pendulum, and (c) angle of inclination for a fixed distance between the straw and the disk of the pendulum.

(a) The first quantity is the minimum distance for which the attraction begins to be detected, as shown by the initial motion of the pendulum due to the approach of the straw. This minimum distance is greater for the attraction between oppositely charged bodies than the minimum distance for the attraction between a charged body and a neutral body. (b) The second quantity is the angle of inclination of a pendulum relative to the vertical for the same distance between the body and the vertical passing through the point of support for the thread of the pendulum. Once more we observe that this angle is greater for the attraction between oppositely charged bodies than the angle for the attraction between a charged and a neutral body. (c) The third quantity the angle of inclination of the pendulum relative to the vertical, considering the same distance between the disk and the straw. This angle is greater for the attraction between oppositely charged bodies than the angle for the attraction between a charged body and a neutral body (Figure 5.6). These three quantities show that this attractive force is clearly much greater between oppositely charged bodies, than the attractive force between a charged body and a neutral body. Thus we conclude that the intensity depends on whether the object brought close to the pendulum is neutral or charged.

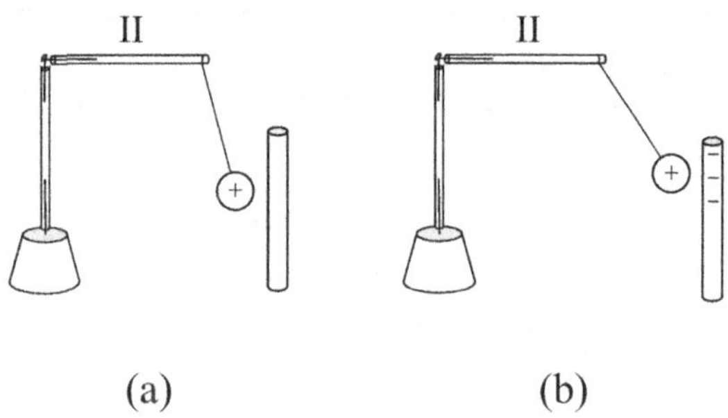

Figure 5.6: (a) The attraction between a positive pendulum and a neutral straw is smaller than the attraction between a positive pendulum and a negative straw, (b).

Similarly, when we slowly bring the rubbed stocking near the negatively charged pendulum I, without allowing them to contact, we observe a greater attraction between them than the attraction between a neutral stocking and the negatively charged pendulum I (Figure 5.7).

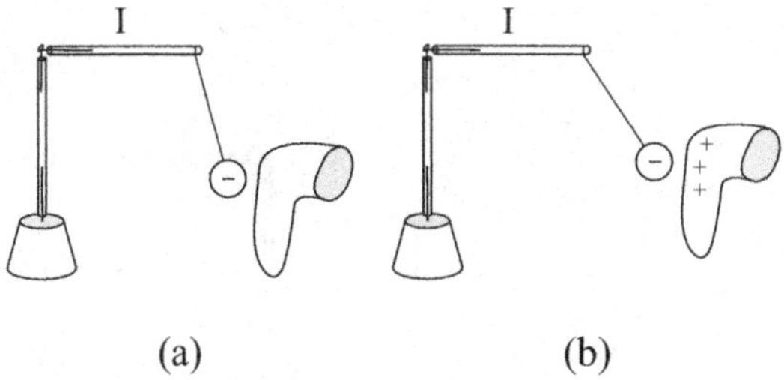

Figure 5.7: (a) The attraction between a neutral stocking and a negative pendulum is smaller than the attraction between a negative pendulum and a positive stocking, (b).

Experiment 5.5

It is also possible to observe another distinction between neutral bodies and charged bodies. Suppose pendulum I is negatively charged and pendulum II is positively charged, as in Experiment 5.1. There is a repulsion when a negatively charged body is slowly brought near pendulum I, while there is an attraction when this body is slowly moved near pendulum II, see Figure 5.8 (a). The opposite happens when a positively charged body is slowly brought near these pendulums. On the other hand, there is an attraction when we move a neutral body near either pendulum I or pendulum II. See Figure 5.8 (b). Sometimes this attraction is so small that it is difficult to detect it.

Experiment 5.6

We now perform some experiments analogous to Experiment 4.4. They are more easily performed when two people work together. They require two plastic straws, two acrylic cloths (see Experiment 5.1) and two pieces of cotton. Each small amount of cotton should take about 10 seconds to fall 2 meters in air. Dandelion seeds can also be conveniently used instead of pieces of cotton. The two straws should be rubbed well with an acrylic cloth, so they become negatively charged. Each person holds one straw horizontally by one of its ends. With the other hand each person releases the cotton a little above his straw. The cotton is attracted by the straw and sticks to it. But then the threads are pushed outward, being repelled by the straw. Sometimes the cotton springs upward and begins to fall in air. When this does not happen, we should slowly blow the cotton until it is released from the straw. We can then keep it floating in air by continually moving the rubbed straw below the cotton, as it is repelled by the straw. In this situation the cotton and the straw are negatively charged.

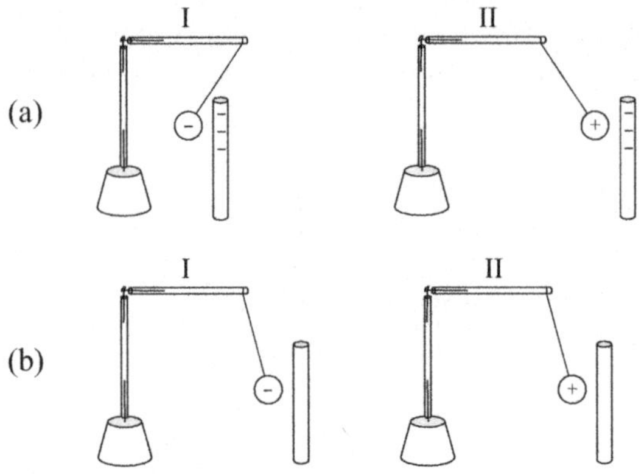

Figure 5.8: (a) A negative body (the rubbed straw) strongly repels another negative body (the paper disk of pendulum I) and strongly attracts a positive body (the paper disk of pendulum II). (b) A neutral body (the unrubbed straw) attracts negative and positive bodies (the paper disks of pendulums I and II). Moreover, the intensities of these forces in cases (a) are greater than those in cases (b).

The new experiment can now begin. Both people are keeping their pieces of cotton floating in air above their rubbed straws. They should now try to direct their pieces of cotton toward one another, trying to bring them into contact in the air. However, they never touch one another, no matter how hard we try. They never get close enough to make contact. It is easy to understand this fact utilizing Du Fay's *ACR* principle. Each floating piece of cotton is repelled by the straw below it as they both have charges of the same sign (negative in this case). Because both pieces of cotton are negatively charged, they repel one another. As a result, we cannot make them touch one another, no matter how hard we try (Figure 5.9).

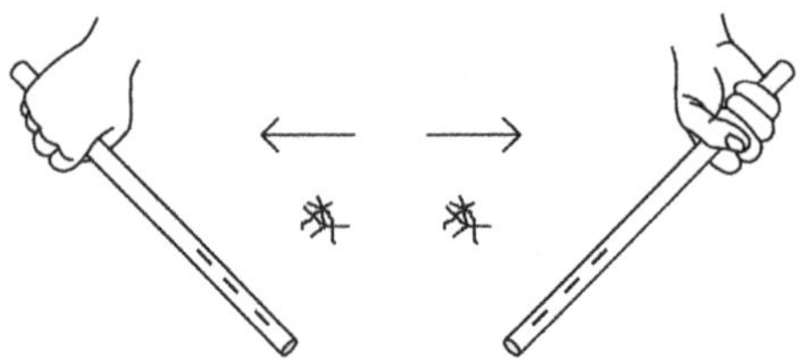

Figure 5.9: It is not possible to join two negatively charged pieces of cotton.

Experiment 5.7

In this experiment we again use two plastic straws that have been negatively charged by being rubbed with an acrylic cloth, as in Experiment 5.6. But now we utilize only a single piece of cotton. Initially we can float the cotton above the straw that has been rubbed with an acrylic cloth, due to the *ACR* mechanism, as in Experiment 4.4. At this moment we bring the second negative straw horizontally near the floating cotton. We observe the cotton moving away from this second straw, as it is also repelled (Figure 5.10).

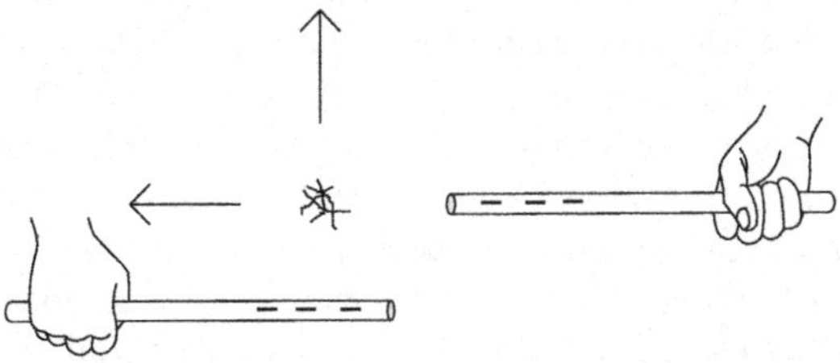

Figure 5.10: Electric forces acting upon a negatively charged piece of cotton.

Experiment 5.8

We repeat Experiment 5.7, initially keeping the negative piece of cotton floating above a negatively charged plastic straw. This time we bring a positively charged glass cup (that is, previously warmed and rubbed with an acrylic cloth) laterally near the piece of cotton. In this case the piece of cotton is attracted by the rubbed glass, moving toward it (Figure 5.11). It is best to make a slow approach, avoiding contact between the cotton and the glass, so as to prevent the piece of cotton from becoming charged by the *ACR* mechanism again, this time positively.

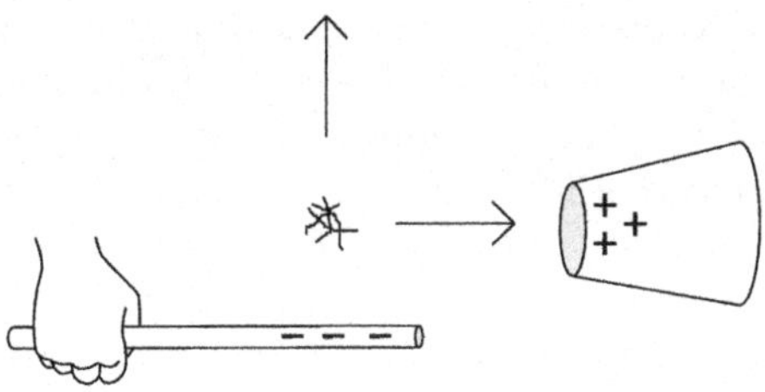

Figure 5.11: A negatively charged piece of cotton is attracted by a positively charged glass.

Experiment 5.9

We now use a piece of cotton, two acrylic cloths, a plastic straw, and a glass cup. This experiment should be performed by two people. However, with practice one person can also perform it. The straw and the glass cup will be rubbed with an acrylic cloth. We know that in this case the straw will become negative and the glass positive. To create a strong charge on the glass it is important to warm it before rubbing, as previously mentioned.

The beginning of this activity is identical to Experiment 4.4. That is, we rub the glass against the acrylic cloth and we hold the glass by its unrubbed portion. The piece of cotton is released above it. The cotton is attracted by the rubbed portion of the glass, touches it, and its fibres are then stretched outwards. Sometimes the cotton jumps outside the glass after a few seconds. If this does not happen, we can again blow the cotton lightly until it is released from the glass. If we bring the glass below the cotton, we can keep the cotton floating above the glass. Sometimes this does not happen at once, so that it is attracted one or more times by the glass until it acquires enough charge to be kept floating above it. The more electrified the glass, the sooner it will keep the piece of cotton floating above it. From now we will suppose this portion of the experiment to be completed. In this case the glass and the cotton floating above it are both positively charged (Figure 5.12 (a)).

While the cotton is floating above the rubbed glass, we slowly bring a negatively charged plastic straw near the cotton, approaching it from above. In this case what is observed is that the cotton is attracted by the straw, in contrast with what happened in Experiment 5.7. Ideally the cotton should not touch the plastic straw. That is, whenever the cotton moves toward the straw, the straw should be moved away from it. With enough practice we can keep the cotton floating between the glass below and the straw above (Figure 5.12 (b)).

In this case it is even possible to remove the glass, so that the cotton will remain floating due only to the attraction of the charged straw above it! In this situation we have the opposite of Experiment 4.4. In Experiment 4.4 the negative cotton was kept floating due to the repulsion of the straw below it. Now, on the other hand, the positive cotton keeps floating due to the attraction of the straw above it (Figure 5.12 (c)). To keep the cotton floating below the straw it is important to move the straw constantly; it cannot remain at rest relative to the Earth, as this is an unstable equilibrium. When the straw is very close to the cotton, the cotton moves quickly toward it and sticks to it, finishing the experiment. On the other hand, if the straw is at a great distance above the cotton, the cotton begins to fall to the ground. Moreover, the piece of cotton tends to move to one side or the other of the vertical plane passing through the straw. As a result, it is necessary to keep the straw constantly in motion, in such a way that the piece of cotton can follow its motion, but without touching it.

When the cotton touches the upper straw, it sticks to it. Sometimes it falls after a few seconds. We can then keep it floating above the straw, since it has again acquired a charge with the same sign as the charge on the straw. In other cases it will only release from the straw when we blow it. In any event, when it is floating above the straw again, we can reverse the situation. When we bring a positive glass toward the negative cotton from above, the cotton will be kept floating between both bodies: the negative straw below and the positive glass above.

In Experiment 5.16 it will be shown how to perform this experiment more easily using two electrified plastic straws.

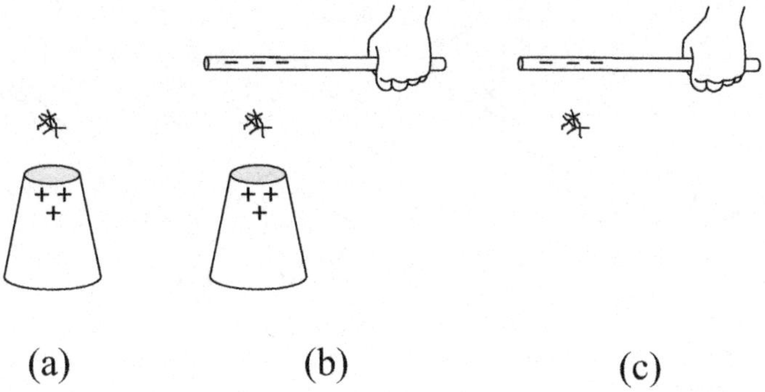

Figure 5.12: (a) A positive piece of cotton can be kept floating above a positive glass. (b) It can also float between a positive glass below and an negative straw above. (c) We can remove the glass and keep the piece of cotton floating below the negative straw.

Experiment 5.10

In this experiment we need two people, a plastic straw, a glass cup, two acrylic cloths, and two small pieces of cotton. One person rubs the glass cup in an acrylic cloth and then keeps the other piece of cotton floating above it. In this case both bodies are positively charged (Figure 5.13 (a)). The other person rubs the straw with an acrylic cloth and then keeps a piece of cotton floating above it. In this case both objects are negatively charged (Figure 5.13 (b)).

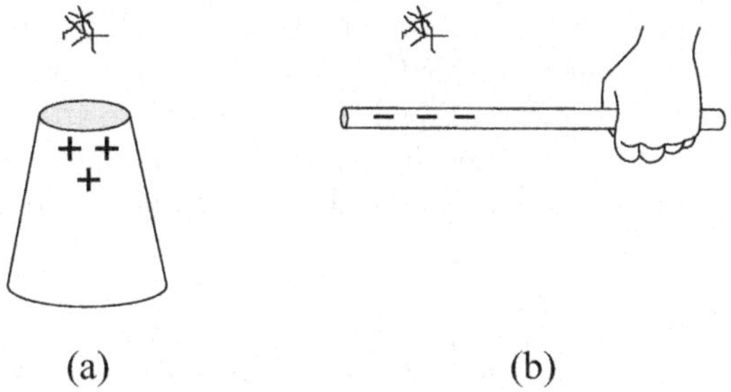

Figure 5.13: (a) A positive piece of cotton floats above a positive glass. (b) A negative piece of cotton floats above a negatively charged plastic.

After this happens, each person tries to direct his piece of cotton toward the other piece of cotton. In this case both pieces of cotton attract one another, stick together, and fall to the ground (Figure 5.14). This is the opposite of what happened in Experiment 5.6.

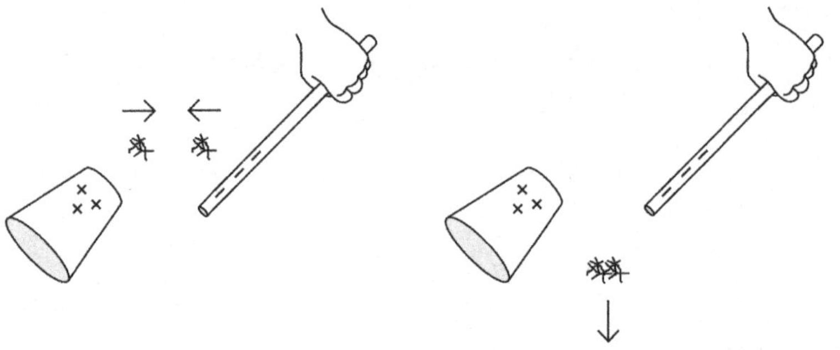

(a) (b)

Figure 5.14: (a) A positive piece of cotton is attracted by a negative one. (b) After touching one another, they fall together to the ground.

5.2 Du Fay Discovers Two Kinds of Electricity

The first to propose the existence of two kinds of electricity was Du Fay in 1733, when he performed experiments similar to the ones just presented. Until that time it was known that electrified bodies attracted light bodies and were attracted by them. Du Fay had recognized another regularity, namely, repulsion between two charged bodies. This was another property of charged bodies, as described in Section 4.1. He had also discovered the *ACR* mechanism, that is, attraction-contact-repulsion. His discovery of two kinds of electricity was completely unexpected. It arose as the result of a curious experimental result which was totally contrary to his expectations. He began by reproducing the experiments with a floating feather done by Guericke, Gray, and Hauksbee. See Experiment 4.4. Initially he electrified a glass tube by friction. He then released small, thin gold leaves above the rubbed tube. They were attracted by the tube, stuck to it, and were then repelled by it. In this way they were kept floating above it. We now quote his words describing the crucial moment of his great discovery (our emphasis in italics):[1]

[1][DF33b, pp. 464-465] and [RR57, p. 586]: "[...] Il demeure donc pour constant, que les corps devenus électriques par communication, sont chassés par ceux qui les ont rendus électriques; mais le sont-ils de même par les autres corps électriques de tous les genres? & les corps électriques ne différent-ils entre-eux que par les divers degrés d'eléctricité? Cet examen m'a conduit à une autre vérité que je n'aurois jamais soupçonnée, & dont je crois que personne n'a encore eu la moindre idée.

J'ai commencé par soûtenir en l'air avec le même tube, deux feuilles d'or, & elles se sont toûjours éloignées l'une de l'autre, quelques efforts que j'aye faits pour les rapprocher, & cela devoit arriver de la sorte, puisque elles étoient toutes deux électriques; mais si-tôt que l'une des deux avoit touché la main ou quelque autre corps, elles se joignoient sur le champ l'une à l'autre, parce que celle-ci ayant perdu son électricité, l'autre l'attiroit & tendoit vers elle: tout cela s'accordoit parfaitement avec mon hypothése, mais ce qui me déconcerta prodigieusement, fut l'expérience suivante.

[...] It is then certain that bodies which have become electrical by communication [that is, by the *ACR* mechanism], are repelled by those which have rendered them electrical. But are they repelled likewise by other electrified bodies of all kinds? And do electrified bodies differ from one another in no respect save their intensity of electrification? *An examination of this matter has led me to a discovery which I should never have foreseen*, and of which I believe no one hitherto has had the least idea.

I began by floating in air with the same [electrified glass] tube, two gold leaves [electrified by the *ACR* mechanism], and they always remained apart from one another, no matter how hard I tried to approach them, and it should happen like this due to the fact that both of them were electrical; but as soon as one of the two [leaves] had touched the hand or any other body, both leaves stuck to one another immediately, due to the fact that the leaf [which had touched the hand], having lost its electricity, the other [electrified leaf] attracted it and moved toward it. [An illustration of this experiment appears in Figure 5.15.] All this was in agreement with my hypothesis, *but what disconcerted me prodigiously*, was the following experiment.

After floating in air a gold leaf by means of the [electrified glass] tube, I approached the leaf with a piece of rubbed and electrified gum copal,[2] the leaf applied itself immediately to it [that is, the gold leaf was attracted by the rubbed gum copal], and stayed there. [An illustration of this experiment appears in Figure 5.16.] *I confess I had expected quite the opposite effect*, since, according to my reasoning, the copal and the gold leaf, which are both electrified, should have repelled each other. [That is, Du Fay expected a repulsion between the two electrified bodies, as he had always observed.] I repeated the experiment many times, believing I had not presented

---

Ayant élevé en l'air une feuille d'or par le moyen du tube, j'en approchai un morceau de gomme copal frottée, & renduë électrique, la feuille fut s'y appliquer sur le champ, & y demeura; j'avouë que je m'attendois à un effet tout contraire, parce que selon mon raisonnement, la copal qui étoit électrique devoit repousser la feuille qui l'étoit aussi; je répétai l'expérience un grand nombre de fois, croyant que je ne présentois pas à la feuille l'endroi qui avoit été frotté, & qu'ainsi elle ne s'y portoit que comme elle auroit fait à mon doigt, ou à tout autre corps, mais ayant pris sur cela mes mesures, de façon à ne me laisser aucun doute, je fus bien convaincu que la copal attiroit la feuille d'or, quoiqu'elle fût repoussée par le tube: la même chose arrivoit en approchant de la feuille d'or un morceau d'ambre, ou de cire d'Espagne frotté.

Après plusieurs autres tentatives qui ne me satisfaisoient aucunement, j'approchai de la feuille d'or chassée par le tube, une boule de cristal de roche frottée & renduë électrique, elle repoussa cette feuille de même que le tube. Un autre tube que je fis présenter à la même feuille la chassa de même, enfin je ne pus pas douter que le verre & le cristal de roche, ne fassent précisement le contraire de la gomme copal, de l'ambre & de la cire d'Espagne, ensorte que la feuille repoussée par les uns, à cause de l'électricité qu'elle avoit contractée, étoit attirée par les autres; cela me fit penser qu'il y avoit peut-être deux genres d'électricité differents, & je fus bien confirmé dans cette idée par les expériences suivantes."

[2]See Appendix A.

the rubbed part of the [copal] rod to the leaf, which accordingly came to the copal as it would to my finger or to any [unelectrified] body; but, having satisfied myself completely on that score, I was entirely convinced that the copal would attract the leaf which the [electrified] tube repelled. The same thing happened by approaching the leaf gold with a piece of [rubbed] amber or with a piece of rubbed Spanish wax.

After several trials which did not satisfy me at all, I approached the gold leaf which had been repelled by the tube, with a rubbed and electrified ball of rock crystal, and it [the ball] repelled this leaf like the tube. Another [electrified glass] tube presented to the leaf likewise repelled it, finally, I could not doubt that glass and rock crystal made exactly the opposite of gum copal, of amber and of Spanish wax, in such a way that the leaf was repelled by the first [rubbed group], due to the electricity it had acquired, being attracted by the second [rubbed group]; *this made me think that there were two different kinds of electricity*, and I was well certified of this idea by the following experiments.

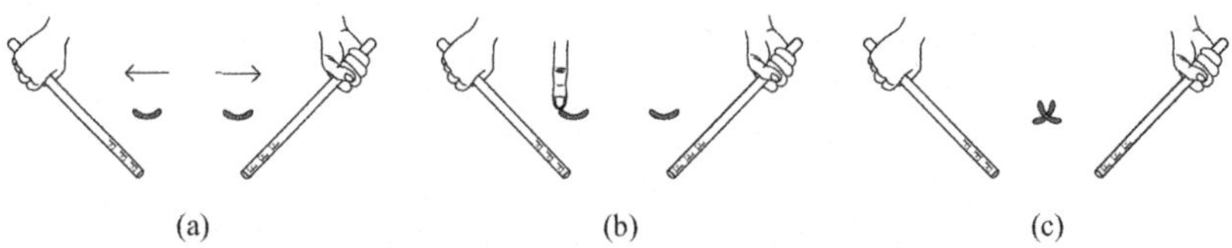

Figure 5.15: (a) Two electrified gold leaves repel one another and the electrified glass tubes. (b) The hand touches one of the floating leaves. (c) After contact with the hand, the leaves approach one another.

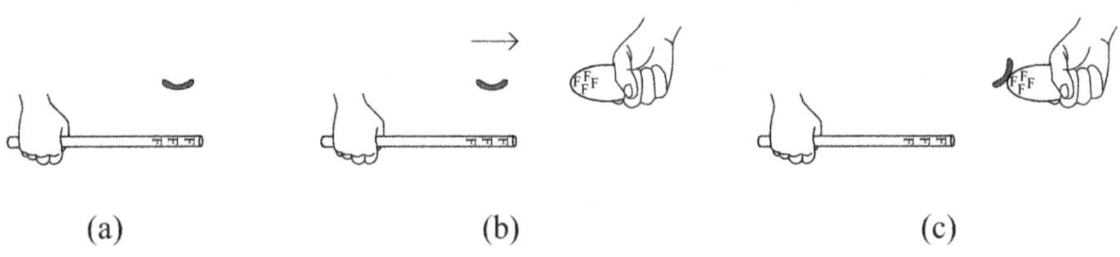

Figure 5.16: (a) An electrified gold leaf floats above a rubbed glass. (b) A rubbed piece of copal moves near the floating leaf. The leaf is attracted by the rubbed copal, with the arrow indicating this new force acting upon it. (c) The leaf moves toward the rubbed copal!

Heilbron gave another English translation of the crucial paragraph of Du Fay's paper mentioned before, the paragraph containing the word *confess*:[3]

[3][Hei99, p. 257].

> I confess I had expected an entirely different effect, because according to my reasoning the copal, being electric, ought to have repelled the leaf, which was likewise electric. I repeated the experiment many times, believing I had not presented the rubbed part of the rod to the leaf, which accordingly came to the copal as it would to any [unelectrified] body; but, having satisfied myself completely on that score, I was entirely convinced that the copal would attract the leaf which the tube repelled.

As the majority of the substances in the first group which he encountered were solid and transparent like glass, he named the first kind of electricity *vitreous electricity*. And due to the fact that the majority of the substances in the second group which he found were bituminous or resinous, he called this second kind of electricity *resinous electricity*:[4]

> We have then two electricities of a totally different nature, namely, the electricity of transparent and solid bodies, like glass, crystal, etc. and the electricity of bituminous and resinous bodies, like amber, gum copal, Spanish wax, etc. Both kinds repel the bodies which acquired an electricity of the same nature as that of their own and which, on the contrary, attract the bodies having an electricity of a different nature than their own.
>
> [...]
>
> Therefore, there are here two electricities well demonstrated, and I cannot avoid giving them different names to avoid confusion of the terms, or the trouble of defining at each moment the electricity I wish to mention. For this reason, I will call one of them *vitreous electricity*, and the other *resinous electricity*, not that I think that there are only bodies having the vitreous nature which are endowed of one [species of electricity], and the resinous matters of the other [kind of electricity], as I have strong proofs of the contrary [point of view], but [I choose these denominations] because the glass and the copal were the two substances which gave me the connections in order to discover the two different electricities.

[4][DF33b, pp. 467 and 469]: "Voilá donc constamment deux électricités d'une nature toute différente, sçavoir, celle des corps transparents & solides, comme le verre, le cristal, etc. & celle des corps bitumineux ou résineux, comme l'ambre, la gomme copal, la cire d'Espagne, etc. Les uns & les autres repoussent les corps qui ont contracté une électricité de même nature que la leur, & ils attirent, au contraire, ceux dont l'électricité est d'une nature différente de la leur. [...]

Voilà donc deux électricités bien démontrées, & je ne puis me dispenser de leur donner des noms différens pour éviter la confusion des termes, ou l'embarras de définir à chaque instant celle dont je voudrai parler; j'appellerai donc l'une l'*électricité vitrée*, & l'autre l'*électricité résineuse*, non que je pense qu'il n'y a que les corps de la nature du verre qui soient doués de l'une, & les matiéres résineuses de l'autre, car j'ai déja de fortes preuves du contraire, mais c'est parce que le verre & la copal sont les deux matiéres qui m'ont donné lien de découvrir ces deux différentes électricités."

Du Fay did not specify which material he used to rub the glass tube and the other substances. He probably rubbed these substances with a cloth made of wool, silk, or cotton.

In a slightly later work he described this casual discovery in the following words:[5]

> Chance has thrown in my way another principle, more universal and remarkable than the preceding one [the *ACR* mechanism, see Section 4.8], and which casts a new light on the subject of electricity. This principle is, that there are two distinct electricities, very different from one another; one of which I call *vitreous electricity*, and the other *resinous electricity*. The first [electricity] is that of [rubbed] glass, rock-crystal, precious stones, hair of animals, wool, and many other bodies. The second is that of [rubbed] amber, copal, gum-lack, silk, thread, paper, and a vast number of other substances. The characteristick of these two electricities is, that a body of the *vitreous electricity*, for example, repels all such as are of the same electricity; and on the contrary, attracts all those of the *resinous electricity*; so that the [glass] tube, made electrical [by rubbing], will repel glass, crystal, hair of animals, etc. when render'd electrick [by rubbing or by the *ACR* mechanism by getting in touch with the rubbed glass tube] and will attract silk, thread, paper, etc. though render'd electrical likewise [by rubbing or by the *ACR* mechanism by getting in touch with a rubbed copal]. [rubbed] amber on the contrary will attract electrick glass, and other [electrified] substances of the same class, and will repel [rubbed] gum-lac, copal, silk, thread, etc. Two silk ribbons rendered electrical [by rubbing], will repel each other; two [electrified] woollen threads will do the like; but a [rubbed] woollen thread and a [rubbed] silk thread will mutually attract one another. This principle very naturally explains, why the ends of threads, of silk, or wool, recede from one another in form of a pencil or broom, when they have acquired an electrick quality. From this principle one may with the same ease deduce the explanation of a great number of other *phenomena*. And 'tis probable, that this truth will lead us to the further discovery of many other things.

As we will see later on, we no longer use Du Fay's terminology. Instead of the terms vitreous and resinous electricities we utilize *positive and negative electricities*, respectively. Other similar expressions used nowadays are *positive and negative electric charges*, or *positively and negatively charged bodies*. Despite this different terminology, Du Fay's fundamental assumption about the existence of two kinds of electricity is still accepted. In modern practice it is also still accepted that charges of the same kind repel one another, while charges of different kinds attract one another.

[5] [DF, pp. 263-264].

A very interesting video showing a modern reproduction of Du Fay's crucial experiment has been made by Blondel and Wolff,[6] "La danse des feuilles d'or."

5.3 Which Kind of Charge does a Body Acquire by Friction?

In the experiments of Chapter 2 we analyzed which substances were or were not attracted by a rubbed object. We also found rubbed substances that had the power of attracting light bodies. Here we will vary the substance with which the objects are rubbed more systematically.

We will use a very practical instrument made of a thin flexible plastic strip attached to a horizontal support (like a pencil, pen, skewer, or straw). The strip can be, for instance, 5 cm wide and 15 cm long. One of its ends is attached to a pencil with adhesive tape. The pencil is kept horizontal and the strip vertical. In Figure 5.17 we see an instrument like this, in profile and face on. By analogy with Gray's pendulous thread of Section 4.9, we can call this instrument a *pendulous plastic thread*, or a *pendulous plastic strip*.

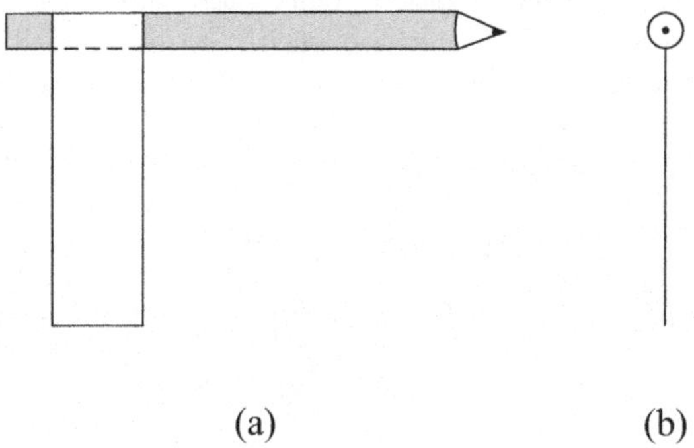

Figure 5.17: A thin flexible plastic strip connected to a pencil. (a) Side view. (b) End view.

Several of these instruments can be made with the plastic taken from the same material (for instance, with all strips cut from the same plastic bag). We should avoid manipulating the strips to prevent them from being charged by friction. Before beginning the experiments with these instruments we should check that they are really neutral. We first discharge a metal versorium by touching it with our finger. We then slowly bring each pendulous plastic strip near the versorium, without allowing them to contact one another. If the versorium does not orient toward the strip, we can consider the plastic to be neutral. When the versorium is oriented by the strip, this plastic should be discarded and we should build another instrument to replace it.

[6][BWa] and [BWb].

Experiment 5.11

We rub two of these neutral pieces of plastic with our fingers, pressing the strip between the forefinger and middle finger, and then quickly moving the fingers downward along the plastic. After this, we bring one horizontal pencil laterally near the other. The pencils can even touch one another. We observe that the strips repel one another due to mutual repulsion. If there were no repulsion they would remain vertically side by side. This is analogous to Experiment 4.1.

We now take two other neutral pendulous plastic strips. We rub them by pressing each strip between two neutral tubes of PVC, like water pipes. To do this, we first check whether these tubes are really neutral. This can be done by bringing each tube near a metal versorium and observing that the versorium is not oriented by the PVC tube. The upper section of the plastic strip is well compressed between two of these neutral PVC tubes. We then pull the pencil with its strip quickly upward. We can check that the strip has become charged after this procedure, by bringing the strip near a versorium, which then turns toward the rubbed plastic. This rubbing procedure should be repeated with the second neutral pendulous plastic strip. We now bring these two strips which were rubbed with PVC tubes together. Once more they repel one another.

There is also repulsion between two plastic strips, initially neutral, which have been rubbed with hair. The same happens for two neutral strips after they have been rubbed between two hard rubber tubes.

These four cases of repulsion are represented in Figure 5.18.

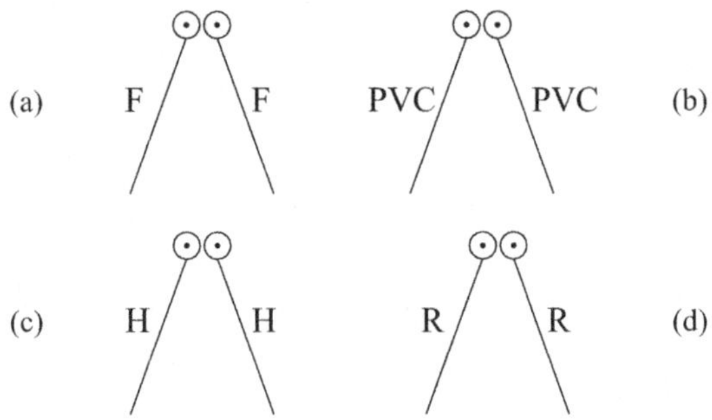

Figure 5.18: Repulsions between two plastic strips rubbed between fingers (F), between two PVC tubes, in hair (H), or between two rubber hoses (R).

Experiment 5.12

We now take one of the plastic strips that have been rubbed between fingers. We bring it close to another plastic strip that has been rubbed with PVC tubes. We observe that they attract one another.

There is also attraction when we bring a plastic strip rubbed between fingers close to another plastic strip that has been rubbed between rubber hoses. The same happens between a plastic rubbed with hair and a plastic that has been rubbed between PVC tubes; or between a plastic that has been rubbed with hair and a plastic rubbed between rubber hoses.

These four cases of attraction are represented in Figure 5.19.

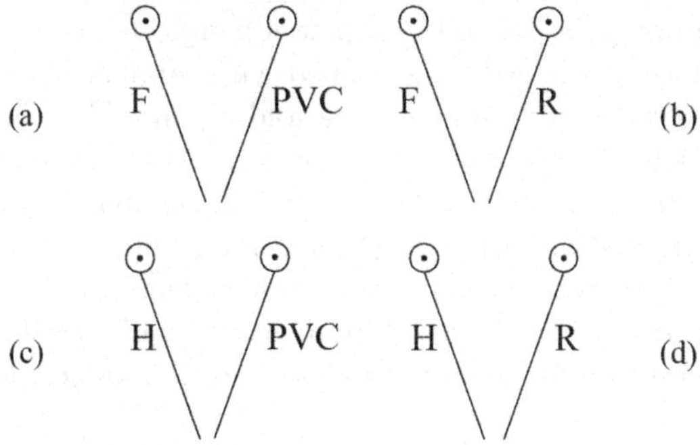

Figure 5.19: Four cases of attraction between two plastic strips rubbed with different substances. (a) Finger and PVC. (b) Finger and rubber. (c) Hair and PVC. (d) Hair and rubber.

Experiment 5.13

On the other hand, there is repulsion between a plastic that has been rubbed between our fingers and another plastic rubbed with hair. The same happens between a plastic rubbed between PVC tubes and another rubbed between rubber hoses (Figure 5.20).

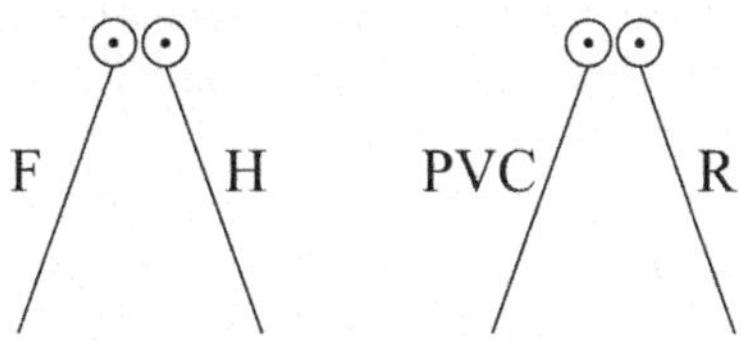

(a) (b)

Figure 5.20: Repulsion between two plastic strips rubbed with different substances. (a) Fingers and hair. (b) PVC and rubber.

These experiments show that a single substance, in our case a plastic strip, can be either negatively charged or positively charged, depending upon the material with which it is rubbed. This indicates that Du Fay's idea of two kinds of electricity seems valid. On the other hand, his supposition that each kind of electricity is associated with a specific group of substances is not valid. Instead of talking of vitreous and resinous electricities, as suggested Du Fay, we adopt the terminology of *positive and negative charges.* In Experiment 5.1 we had separated the charged bodies into two distinct groups. In our example, the first group was composed of a plastic straw rubbed with an acrylic cloth, an acrylic ruler rubbed with an acrylic cloth, a pendulum charged by contact with this plastic straw, and a pendulum charged by contact with this rubbed acrylic ruler. In our example, the second group was composed of a glass cup rubbed with an acrylic cloth, a silk stocking rubbed with an acrylic cloth, a pendulum charged by contact with this rubbed glass, and a pendulum charged by contact with this rubbed silk. It was observed that objects in the first group repelled one another, objects in the second group repelled one another, and an object in the first group attracted an object in the second group.

Convention: The convention now is that objects in the first group have become *negatively charged*, or have acquired *negative charge.* It is also said that objects in the second group have become *positively charged*, or have acquired *positive charge.*

Du Fay himself tried to determine whether the kind of electricity acquired by an object might depend upon the material with which it was rubbed.[7] In order to test this influence he rubbed a warmed silk cloth first with his hands and then with another warmed silk cloth. He verified that the rubbed silk acquired the same resinous electricity in both cases. He also rubbed wool and feathers with his hands and with silk. In both cases he found that the wool and the feathers acquired the same vitreous electricity. After these few tests he concluded that the material with which we rub a body could change the amount of electrification acquired by the body, but not the kind of electricity it acquired. Later on this conclusion had to be modified when it was found that the same body could acquire both kinds of electric charge, depending upon the material with which it was rubbed. It therefore no longer makes sense to talk of *vitreous electricity* or *resinous electricity*, since glass itself, for instance, can acquire both kinds of electricity, depending upon the substance with which it is rubbed.

Definitions: Nowadays we talk of *positive and negative electricities*, or *positive and negative charges.* Moreover, it is conventional to call the charge acquired by a plastic straw rubbed with hair, skin, cotton, or silk *negative.* Likewise, the charge acquired by a plastic straw rubbed with hard rubber, acrylic tube, or PVC is called *positive.*

Experiment 5.14

In order to determine which charges were acquired by the rubbed plastics in Experiment 5.11, we do an experiment that begins much like Experiment 5.1. An electric pendulum I is negatively charged by contact with a plastic straw charged by friction with an acrylic cloth. An electric pendulum II is positively charged by contact with a silk stocking charged by friction with an acrylic cloth. We place these two pendulums at a good distance from one another. The rubbed straw and the rubbed stocking are removed from the table. The threads of the

[7][DF33b, pp. 472-473].

two charged pendulums hang vertically. We rub a pendulous plastic strip with our fingers. We slowly bring this rubbed plastic near pendulum I, not allowing them to contact one another. We observe that the pendulum is repelled by the plastic (Figure 5.21 (a)). After this procedure, we slowly bring this rubbed plastic near pendulum II, not allowing them to contact one another. We observe an attraction between them (Figure 5.21 (b)). We thus conclude that the plastic we rubbed with our fingers has become negatively charged.

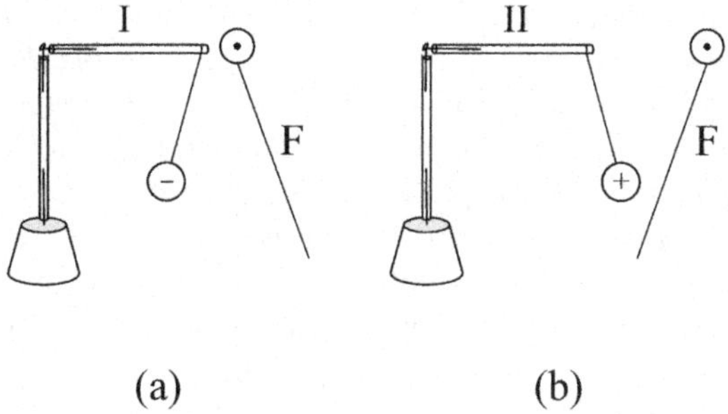

Figure 5.21: Du Fay's procedure for discovering the sign of the charge of an electrified body. In this example the rubbed plastic is being repelled by a negatively charged pendulum, (a), and is being attracted by a positively charged pendulum, (b). Therefore, we conclude that the plastic is negatively charged.

By repeating this experiment with other plastic strips rubbed with different substances, we observe that it becomes charged as follows: between fingers (negatively), between PVC tubes (positively), in hair (negatively), between hard rubber hoses (positively).

This is the procedure for determining which kind of charge a body has acquired by being rubbed against a certain substance. Essentially we need to know beforehand that a body I is negatively charged and that a body II is positively charged. We then bring the charged test body near body I and also bring it near body II. When the test body is repelled by I and attracted by II, it is said that it has a negative charge. When it is attracted by I and repelled by II, it is said that it has a positive charge. This procedure is due to Du Fay, although he utilized the concepts of vitreous and resinous electricities, instead of our positive and negative electricities:[8]

> In order to judge the kind of electricity of any body, it is only necessary to electrify it and to present it to an [electrified] piece of glass and then to an [electrified] piece of amber; it [the body] will be constantly attracted by one [piece] and repelled by the other [piece]; [...]

[8][DF33b, pp. 469-470]: "Pour juger donc quelle est l'espece d'électricité d'un corps quelconque, il n'y a qu'a le rendre électrique, & lui présenter l'un après l'autre un morceau de verre & un morceau d'ambre, il sera certainement attiré par l'un, & repoussé par l'autre; [...]"

We will perform this procedure more systematically in Section 5.4.

Experiment 5.15

We now know that any substance can become positively or negatively charged, depending upon the substance used to rub it. We can thus repeat Experiment 5.1 in a more practical way, yielding more visible results.

A plastic straw acquires a good amount of negative charge when rubbed with hair, in the skin, or with cotton. By trial and error it is found that it acquires a good amount of positive charge by being rubbed between two hard rubber hoses. To charge the straw, we cut two pieces of this hose. We place one end of the straw between these well compressed pieces and pull it quickly away from the rubber hoses. A plastic straw also acquires a good amount of positive charge by being rubbed in this way between two PVC tubes.

We charge the paper disk of pendulum I by the ACR mechanism through contact with a plastic straw rubbed with hair. After contact, this pendulum is repelled by this straw. We charge the paper disk of pendulum II by contact with a plastic straw rubbed against two hard rubber hoses. After contact, this pendulum is repelled by this straw.

However, when we slowly bring this last rubbed straw near pendulum I, without allowing them to contact, there will be an attraction between them. Likewise, when we slowly bring the first rubbed straw near pendulum II, without allowing them to contact, there will be an attraction between them.

The advantage of this procedure over the one adopted in Experiment 5.1 is that it is easier and more practical to acquire a good amount of positive charge with a plastic straw rubbed between two hard rubber hoses, than with a glass cup (or a silk cloth) rubbed against an acrylic cloth. In particular, it is difficult to electrify the modern glasses by friction while holding the glass cup in our hand.

Experiment 5.16

Experiment 5.9 can be repeated more easily using two plastic straws rubbed with different materials (since the same object rubbed with different materials will acquire different charges), instead of using a straw and a glass cup rubbed with the same material. We use also a dandelion seed, or a piece of cotton which takes some 10 seconds to fall a distance of 2 meters. We rub a straw in hair to charge it negatively. We rub another straw against two hard rubber hoses to charge it positively, as in Experiment 5.15. These two straws are held horizontally but separated from one another.

The dandelion seed is released above the negative straw. It is attracted by the straw, touches it, and is then repelled by it (if it does not jump straight away from the straw, we can blow it gently). The dandelion seed acquires a negative charge by the ACR mechanism. It can then be kept floating above the straw, as in Experiment 4.4 (Figure 5.22 (a)).

We now slowly bring the positive straw above the dandelion seed. The second straw should always remain above the dandelion seed, without touching

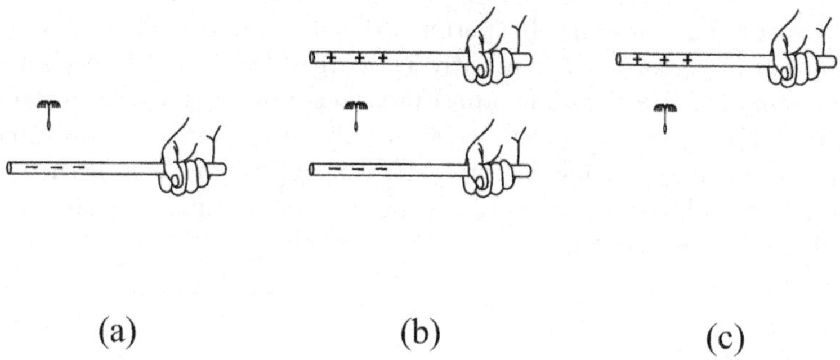

Figure 5.22: (a) A negative dandelion seed floating above a negative straw. (b) The negative dandelion seed floating between a negative and a positive straw. (c) The negative dandelion seed floating below a positive straw.

it. In this case there will be an attraction between the negative dandelion seed and the second positive straw. The negative dandelion seed can remain floating between the lower negative straw and the upper positive straw, as in Figure 5.22 (b). This can only be obtained by avoiding contact between the negative dandelion seed and the positive straw above it.

It is possible to remove the lower negative straw and keep the negative dandelion seed floating in air below the positive straw, provided they do not touch one another (Figure 5.22 (c)). With practice we acquire a good control of this experiment. Whenever the negative dandelion seed moves toward the upper positive straw, we must remove the straw quickly, but always maintaining it above the dandelion seed. In this way the dandelion seed can be kept floating below the straw, while oscillating up and down below it.

Experiment 5.17

Experiment 5.15 can also be performed utilizing the Du Fay versorium, Section 4.7. We build two of these versoria, I and II, always checking whether they are really discharged after we build them. Recall the use of aluminum foil on the end of one leg of each Du Fay versorium.

Initially we rub a plastic straw in hair to charge it negatively. We then charge versorium I by the ACR mechanism, as described in Experiment 4.12. After the aluminum foil is released from the negative straw, it is repelled by it. We rub another straw between two hard rubber hoses in order to charge it positively. We charge versorium II positively by contact with this straw. After the aluminum foil releases this straw, it is repelled by it.

We now slowly bring the negative straw near versorium II, not allowing them to contact one another. We observe that the aluminum foil is attracted by the straw. We slowly bring the positive straw near versorium I, not allowing them to contact one another. We observe that the aluminum foil is also attracted by this straw.

Experiment 5.18

Experiment 5.14 can also be performed with two Du Fay versoria. The aluminum foil of versorium I is negatively charged by the ACR mechanism, as in Experiment 5.17. By the same procedure the aluminum foil of versorium II is positively charged. Another plastic straw is rubbed against a certain substance. After this procedure, we bring the rubbed section of this straw near the negative versorium I, not allowing the two to ocontact. Let us suppose that they repel one another, as in Figure 5.23.

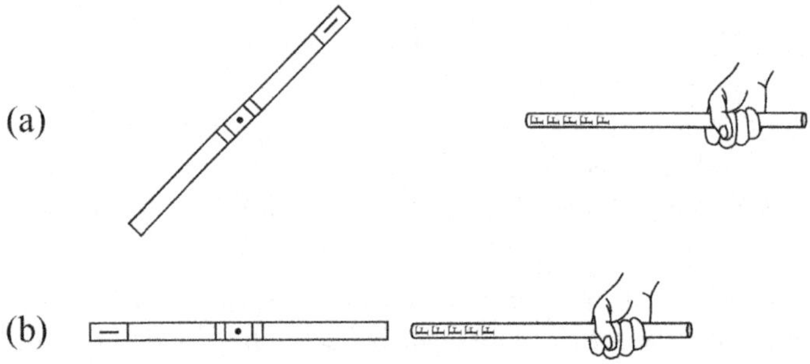

Figure 5.23: Negatively charged aluminum foil of a Du Fay versorium being repelled by an electrified body approaching it.

We now move the rubbed portion of this straw near the positive versorium II, not allowing them to contact one another. Let us suppose that they attract one another, as in Figure 5.24. As like charges repel one another, and opposite charges attract one another, in this case we conclude that the straw has become negatively charged by friction. The same procedure can be utilized in order to determine the sign of the charge on other materials electrified by friction against different substances.

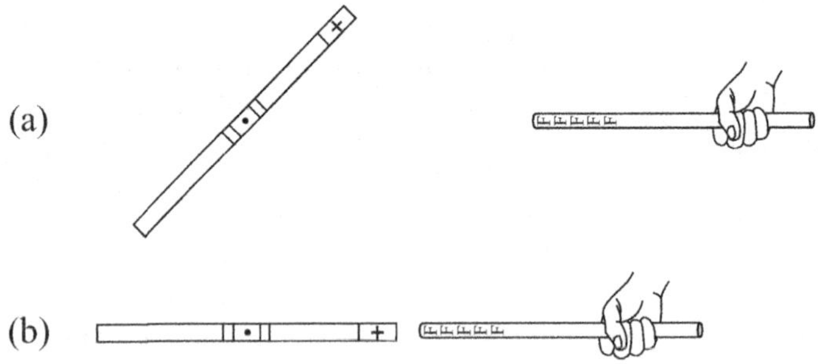

Figure 5.24: Positively charged aluminum foil of a Du Fay versorium being attracted by an electrified body brought near it.

Du Fay created this kind of versorium as a practical and sensitive instrument to determine the kind of electricity acquired by objects that take on very little electricity. Initially he built a metal versorium, but later on he noticed that, for his purposes, it would be better to have versoria made of glass or wax. He then described an experiment analogous to Experiment 5.18:[9]

> There is still another very simple way to know the kind of electricity of a body for which this virtue is very weak [...] For the time being I will avoid [presenting] the reader with the tiresome and discouraging details of the faulty or imperfect experiments, and I will only say that in order to succeed, it is necessary to utilize a glass needle [a versorium made of glass] placed above a very long glass pivot, this needle should have in one of its ends a hollow metal ball, and in the other [end] a counter-weight of glass, it is necessary to dry all its portions, and then it is necessary to communicate the electricity to the metal ball with a [rubbed glass] tube, or of any analogous matter, the [charged] ball will be then attracted by bodies which have a resinous electricity, and will be repelled by the bodies which have a vitreous electricity.

5.4 The Triboelectric Series

In this experiment we illustrate how the practical procedure of Experiment 5.14 can be utilized with a wide variety of substances rubbed against different materials. A metal versorium will be utilized to test whether the body is neutral or charged. Initially we touch the versorium with our finger, in order to discharge it. We then slowly move the body near the versorium, without touching it. When the versorium remains at rest, without being oriented toward the body, we say the body is neutral. When the versorium is oriented toward the body, we say the body is charged.

In order to test the charge acquired by rubbed bodies, we utilize two electric pendulums charged with opposite electricities. These two pendulums are kept at a good distance from one another on the table. Before charging the pendulums we touch the paper disk of each one of them with our finger. We then charge them by contact. A simple and efficient procedure is to rub a plastic straw and bring it close to one of the pendulums. The paper disk is attracted by the rubbed straw, touches it, and is then repelled by it, as it acquires a charge of the same sign.

From our experience with this kind of experiment, we know that it is easy to charge a pendulum negatively. After all, a plastic straw acquires a large amount

[9] [DF33b, pp. 473-474]: "Il y a encore un moyen bien simple pour connoitre le genre d'électricité d'un corps dans lequel cette vertu est très-faible [...] J'épargnerai cependant au lecteur un détail ennuyeux & rebutant d'expériences manquées ou imparfaites, & je dirai seulement que pour réussir, il faut se servir d'une aiguille de verre posée sur un pivot de verre très-long, que cette aiguille porte à l'un de ses bouts une boule de métal creuse, & à l'autre un contre-poids de verre, qu'il faut bien sécher toutes ces piéces, & qu'alors il faut communiquer l'électricité à la boule de métal avec le tube, ou quelqu'autre matiére analogue, la boule sera alors attirée par les corps dont l'électricité est résineuse, & repoussée par ceux qui ont l'électricité vitrée."

of negative charge by being rubbed against several substances: hair, cotton, paper, etc. There are other bodies which also acquire a good amount of negative electricity (the substance with which they are rubbed between parentheses): acrylic tube (sheet of paper, cotton, acrylic cloth, hair, synthetic polyamide), hard rubber hose (plastic bag, acrylic cloth, synthetic polyamide, hair, glass), PVC tube (human skin, acrylic cloth, synthetic polyamide, glass), Styrofoam (hair), plastic bag (hair), and silk (hair).

It is more difficult to find appropriate substances which acquire a large amount of positive electricity by friction. In order to charge a plastic straw with a great amount of positive charge we had to rub it against two hard rubber hoses (or against two PVC tubes), by pulling it quickly between the compressed hoses. Some other objects which acquire a good amount of positive charge (the substance with which they are rubbed between parentheses): glass (acrylic cloth), synthetic polyamide (acrylic cloth), and silk (acrylic cloth, hard rubber hoses, acrylic tubes, and PVC tubes).

In order to charge some substances having the shape of a thread or wire (hair, cotton thread, synthetic polyamide thread, and polyester thread) we utilized the following procedure. They were initially tied to a plastic straw to prevent them from being discharged through our hand after the friction (Figure 5.25 (a)). To test the charge acquired by these rubbed threads we did not employ the charged pendulums described previously. Instead of this, on appropriate supports we stood a negatively charged straw (rubbed with hair along its entire length) and a positively charged straw (rubbed between two hard rubber hoses along its entire length). The support can be identical to the pendulum support (Figure 5.25 (b)).

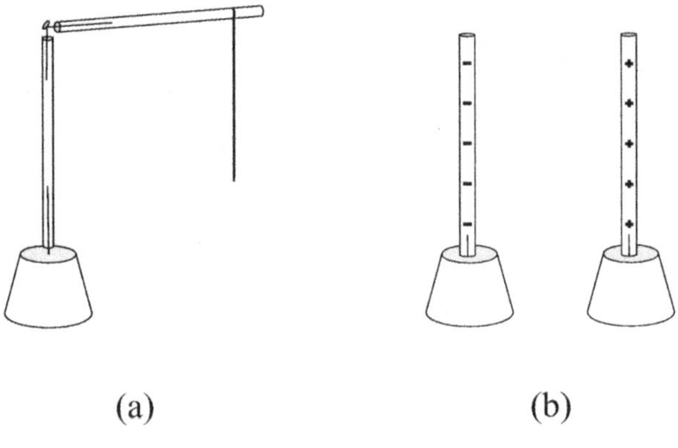

Figure 5.25: (a) Thread tied to a plastic straw. (b) Oppositely electrified plastic straws.

The rubbed thread, tied to a horizontal straw, was then moved slowly near the negatively charged vertical straw, without touching it. Afterwards it was slowly moved near the positively charged vertical straw, without touching it. By observing the attraction and repulsion of this thread in relation to these charged straws, we can determine the charge acquired by friction in this thread.

The objects tested were the following: dry human skin, plastic bag, tissue paper, Styrofoam, aluminum foil, cotton, acrylic cloth (see Experiment 5.1), synthetic polyamide, hair, leather, steel wool, cork, smooth glass, hard rubber hose, drinking plastic straw, porcelain, acrylic tube, PVC tube, thin cardboard, wood, and silk. We rubbed some substances with hydrophilous cotton, or rubbed a cotton thread against other bodies. Synthetic polyamide was utilized in the form of a women's stocking to rub some substances. Some so-called "silk" stockings are in fact made of synthetic polyamide. Synthetic polyamide is also easily found in swimming suits. We rubbed different objects against our hair, but we also used a single human hair tied to a plastic straw and rubbed it against other substances. The wood used in our experiments was obtained from barbecue skewers sold in stores. We are not sure from which kind of tree this wood was obtained. We used a silk cloth to rub some substances. We also utilized a silk thread, which was rubbed against other objects. The glass utilized here was very smooth.

The procedure adopted in most cases was the following. A pendulum was negatively charged by touching a straw rubbed with hair. Another pendulum was positively charged by contact with a straw rubbed against two hard rubber hoses. These charged pendulums were spatially separated far enough apart such that their threads would hang vertically. The rubbed straws were also spatially separated and were kept vertically supported by appropriate supports, in order to test the charge acquired by some rubbed threads.

We first would check whether certain objects I and II were initially discharged by moving each one of them near a versorium. If this was the case, object I would then be rubbed against object II. The idea being that we could vary both objects I and II to determine the charge produced.

After doing this, we checked whether object I had received enough charge. To do this, we slowly brought it near the versorium, to see if the versorium would be oriented by it. If it was, the charged object I would be slowly moved near the negatively charged pendulum, without contact, and it would be observed if the pendulum was attracted or repelled by this object. Object I was then slowly moved near the positively charged pendulum, without contact. We observed whether this second pendulum was attracted or repelled by object I. If there was attraction (repulsion) of the negatively charged pendulum, and also repulsion (attraction) of the positively charged pendulum, we concluded that object I had become positively (negatively) charged by being rubbed against object II.

It seems to us relevant to check whether one of the pendulums was attracted or repelled by the charged object, and whether the other pendulum was repelled or attracted by this charged object. This precaution is necessary in order to prevent any doubt as to the charge acquired by the rubbed object. The reason for this precaution is that the charge acquired by friction can often be very small. This creates attractions and repulsions of small intensities between this rubbed object and the charged pendulums, making precise observations difficult to perform.

From time to time we should touch the versorium with our finger, to discharge any residual electricity it may acquire. Moreover, from time to time we should also charge each pendulum again, as their electricities decrease with the passage of time. As regards the plastic or resinous objects, we should not rub the same object against more than one substance. The reason for this precaution is that sometimes the charge acquired during the first friction remains in the plastic or resinous body. This makes it difficult to know clearly which kind of electricity was acquired by friction with the second substance. In order to test the charge acquired by a plastic straw, for instance, we should utilize a new straw for each substance we will rub it against. Moreover, before rubbing the straw, it is always wise to test if it is really neutral.

In order to rub the plastic straw against hard substances, we fixed one of its ends between them (between two pieces of Styrofoam, between two corks, and so on). After this procedure, we quickly pulled the straw away from these two objects, moving its whole length between them.

In the case of glass, leather, porcelain, wood, and paper we need special care before manipulating them with our hands. They can only acquire and maintain a good amount of charge if, before friction, they have been dried and, preferably, warmed. We heated them in a microwave, but it is also possible to heat them with a match.

In general we rubbed body I against body II. Whenever possible we tested both charges, the charge on body I and the charge on body II. But this was not always possible. In some cases one of these bodies did not maintain any charge generated by friction. Normally this body is discharged by contact with our hand, as is the case with steel wool, cotton, or a sheet of paper. If this was the case, we tested only the charge maintained by the other body.

After following these procedures and performing the experiments, we obtained the result shown in Table 5.1.

Definition: A list like Table 5.1 is called a *triboelectric series*. The prefix "tribo" comes from the Greek. Its meaning is friction or the act of rubbing. A triboelectric series indicates the kinds of electrification obtained by friction.

This Table should be read as follows: When we rub body I against body II, the positively charged one will be the body which is above the other. That is, the body which is closer to the symbol + will become positively charged, while the other becomes negatively charged. For instance, when the plastic straw is rubbed in silk, the silk will become positive and the plastic negative.

We did not include steel wool in this Table because it was difficult for any body to acquire a strong electric charge by being rubbed against it. The bodies which have become negatively charged were the plastic straw, the acrylic tube, the hard rubber, the PVC tube, the Styrofoam, and the plastic bag. A single human hair, on the other hand, has become positively charged by being rubbed against steel wool. The glass, the wood, the porcelain, the acrylic cloth, the synthetic polyamide, and the silk did not acquire a perceptible amount of charge.

This triboelectric series does not coincide with other triboelectric series found in the literature. It is also common to find two or more triboelectric series in the literature which are different from one another, even including the same bodies in different order. There are good reasons for this divergence of results.

+
hair
smooth glass
human skin
synthetic polyamide
cotton
silk
paper or thin cardboard
leather
porcelain
aluminum foil
wood
cork
acrylic cloth
Styrofoam
plastic bag
drinking plastic straw
rigid acrylic
PVC tube
hard rubber
−

Table 5.1: Triboelectric Series.

There are many different kinds of glass, made of different materials and with varying fabrication procedures. These aspects certainly affect their capacity to acquire positive or negative charges by friction. For instances, some kinds of glass became positively charged after being rubbed in our hand, while other kinds of glass became negatively charged. The same variety of materials and fabrication procedures happens with other substances (plastic, paper, rubber, etc.) The dye used in silk and in other cloths or threads can also affect their properties. Wood from different trees can have different properties. Human hair and skin can be more or less oily, can also be impregnated with shampoos, cremes, and other substances.

Everyone who does these experiments should try to build his own triboelectric series, utilizing available materials.

In Section 5.3 we saw how, in 1733, Du Fay had rubbed silk, feathers, and wool against silk and against his hand. He found that each one of these bodies was always charged with electricity of the same kind, no matter which material it was rubbed against. Since then some anomalies were detected. John Canton (1712-1772), in particular, found that roughened (unpolished) glass could be charged positively by being rubbed with flannel, or negatively by being rubbed with oiled silk.[10] These researches were continued by Johan Carl Wilcke (1732-

[10][Can54].

1796), who in 1757 published the first triboelectric series, namely:[11] smooth glass, wool, quills, wood, paper, sealing wax, white wax, rough glass, lead, sulphur, and metals other than lead. In 1759 Benjamin Wilson (1721-1788) published another series, probably obtained independently from Wilcke, namely:[12] diamond, tourmaline, glass, amber. These were the first triboelectric series ever published.

5.5 Are Attractions and Repulsions Equally Frequent?

Experiment 5.19

In Section 5.4 we saw how to charge a plastic straw positively and negatively. We charge one of these straws negatively by friction and repeat Experiment 4.10. That is, initially we touch the paper disk of the pendulum with our finger. We then move the negative straw near the pendulum. The paper disk is attracted by the straw, touches it, and is then repelled by it. When we touch the paper disk with our finger, it discharges. We can then repeat the whole procedure.

The same experiment should be done with a positively charged straw. Initially we discharge the pendulum when we touch the paper disk with our finger. When we bring the positive straw near the disk, the disk is attracted by the straw, touches it, and is then repelled. When we touch the paper disk it discharges. The whole procedure can then be repeated.

Experiment 5.20

We now perform experiments analogous to Experiments 5.1 and 5.15. We charge pendulum *I* negatively by the *ACR* method, as in Experiments 4.7 and 5.19. We charge pendulum *II* positively by the *ACR* method, as in Experiment 5.20. We slowly bring the negative straw near negatively charged pendulum *I*, observing the repulsion of the pendulum. We now move the negative straw near positively charged pendulum *II*, without bringing them into contact. We observe an attraction of the pendulum. We slowly bring the positive straw near positively charged pendulum *II*, observing that the pendulum is repelled. We now move the positive straw near negatively charged pendulum *I*, without bringing them into contact. The pendulum is attracted by the straw.

Experiment 5.21

We now perform experiments analogous to Experiment 4.8. A pendulum is negatively charged by the *ACR* method. To charge it negatively, a straw rubbed with hair is brought near the pendulum, as in Experiments 5.1 and 5.15. After the paper disk is repelled by the rubbed straw, we remove the straw. We

[11] [Hei99, pp. 387-388].

[12] [Wil59] and [Hei99, pp. 387-388].

now slowly bring our finger near the paper disk, without making contact. The pendulum is attracted by the finger.

The same procedure is repeated with a positively charged pendulum. A positive charge can be obtained when we bring a straw rubbed between two hard rubber hoses near a neutral pendulum. After the paper disk is repelled by this rubbed straw, we remove the straw. We now slowly bring our finger near the pendulum, without making contact. Once more the pendulum is attracted by the finger.

Experiment 5.20 again shows that a negatively charged body I repels another negatively charged body II. The same happens for two positively charged bodies. On the other hand, if body I is negatively charged and body II is positively charged, they attract one another. The same happens if I is positive and II negative.

We thus see two attractions and two repulsions. This might indicate that these two phenomena are equally frequent. However, as seen in Experiments 5.19 and 5.21, a charged body normally attracts a neutral body, no matter whether the charged body is positive or negative. And a neutral body like our finger attracts not only a positive body, but also a negative one. This shows that attractions are much more frequent or common than repulsions, due to the fact that the majority of bodies are macroscopically neutral. When we charge a certain body, it will tend to attract almost all bodies around it, although it may happen that this attraction is of very low intensity, difficult to be detected or observed. This charged body will only try to repel the other bodies which have a net charge of the same sign. If the second body has a charge of opposite sign to the first one, or if the second body is neutral, there will be an attraction between them.

In conclusion we may say that these experiments illustrate the fact that electrical attractions are much more common than electrical repulsions, as they occur much more frequently.

In Section 7.10 we will discuss the force of interaction between two bodies electrified with charges of the same sign.

5.6 Variation of the Electric Force as a Function of Distance

Since the time of the oldest experiment in electrostatics, Experiment 2.1, it has been known that the attraction exerted by a charged body upon a small light body depends upon the distance between them. After all, light bodies are only visibly attracted by a rubbed body when the distance between them is small. When the distance between these two bodies is very large, this attraction is not easily detected. The same happens with the versorium experiments, like Experiment 3.1. That is, only when the rubbed plastic is brought near the versorium does the versorium orient toward the plastic. The same happens with all other experiments described so far, as the effects only happen when there is a small distance between the interacting bodies.

These effects can also be observed in attractions and repulsions between charged bodies. Let us illustrate this fact with an electric pendulum.

Experiment 5.22

An electric pendulum *I* is negatively charged by the *ACR* method, as in Experiments 5.1 and 5.15. After this procedure, we remove the negative straw which charged it by contact. In this situation the thread of the pendulum returns to the vertical. We charge another straw positively when we rub it between two hard rubber hoses. We slowly bring the positive straw near the negative pendulum, without contact, until we can detect their attraction, as shown by the inclination of the thread relative to the vertical. The horizontal straw should be at the same height as the paper disk, approaching it from the side. We slowly decrease the distance between the paper disk and the tip of the rubbed straw. We observe that decreasing this distance, increases the angle of inclination of the straw relative to the vertical (Figure 5.26). This shows that the force of attraction between oppositely charged electrified bodies increases when the distance between the interacting bodies decreases. The force is being indicated in this experiment by the angle of inclination of the pendulum in relation to the vertical.

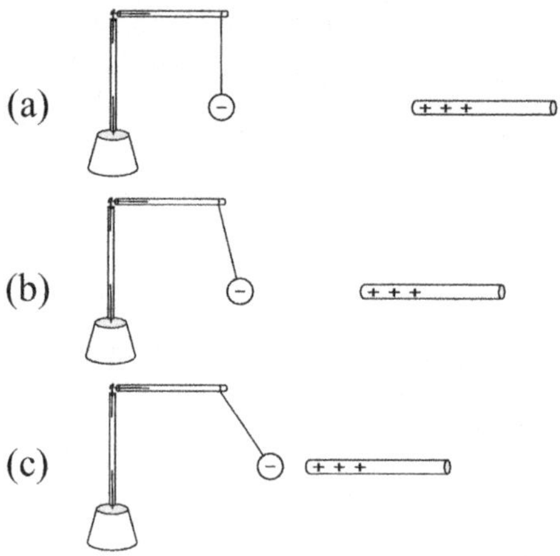

Figure 5.26: By decreasing the distance between the negative paper disk and the positive tip of the rubbed straw, we increase the attractive force between them.

Experiment 5.23

An electric pendulum *I* is negatively charged by the *ACR* method, as in Experiments 5.1 and 5.15. After this procedure, we remove the negative straw which charged it by contact. In this situation the thread of the pendulum returns to the vertical. We now slowly bring the negative rubbed straw near the charged pendulum until the distance at which repulsion begins to be visible, as indicated by the deviation of the pendulum from the vertical. The straw should be kept horizontal and at the same height as the paper disk, and brought toward it by the side. We then slowly decrease the distance between the paper disk and the rubbed tip of the straw. We observe that when this distance is decreased, the angle of inclination of the thread to the vertical increases, as shown in Figure 5.27. And this indicates an increasing force of repulsion with decreasing distance between the interacting bodies.

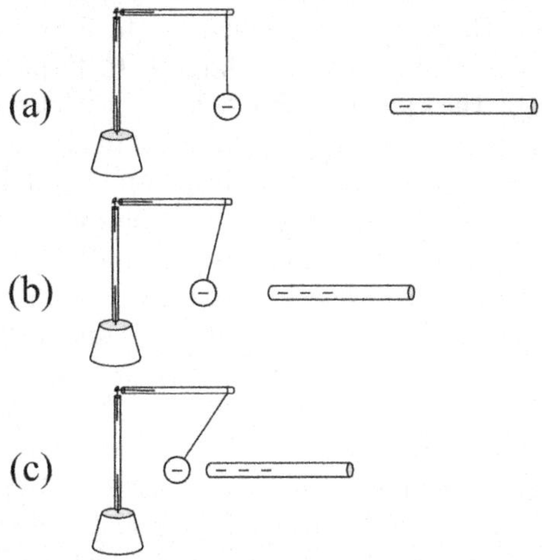

Figure 5.27: By decreasing the distance between the negative paper disk and the negative tip of the rubbed straw, we increase the repulsive force between them.

In this experiment the rubbed straw should not come too close to the electrified pendulum. The reason is that in some situations they can attract one another when the mutual distance between them is too small. This will be discussed in Section 7.10.

5.7 Variation of the Electric Force with the Quantity of Charge

Thus far we have not bothered to measure the quantity of electricity (or the magnitude of electrical charge). Normally this is done by means of the notion of electric force.

Let A, B, and C be three bodies whose sizes are small in comparison with the distance between them. We will consider that bodies A and B are electrified (by friction or by the ACR mechanism). We call F_{AC} the force between A and

C when they are separated by the distance d, with body B far away from these two bodies. We call F_{BC} the force between B and C when these two bodies are separated by the distance d, with A far away from these two bodies.

Definitions: We say that the magnitude of charge A is equal to the magnitude of charge B when $F_{AC} = F_{BC}$. If F_{AC} is bigger than F_{BC}, then we say that the magnitude of charge A is greater than the magnitude of charge B. If F_{AC} is smaller than F_{BC}, then we say that the magnitude of charge A is smaller than the magnitude of charge B.

The intensity of the force can be measured by different means. Here we consider situations where the distances are always the same. In Experiment 2.1, for instance, the more pieces of paper collected by the straw, the greater the force it exerts upon them. In Experiment 2.8 the force intensity is indicated by the curvature of the falling stream of liquid. In Experiment 4.1 the force intensity is indicated by the opening angle between the electrified plastic strips. In the experiments with the electric pendulum, as in Experiment 4.5, the force intensity is indicated by the angle between the pendulum and the vertical. In the experiments with the pendulous plastic thread the force intensity is indicated by the opening angle between the plastic strips, as in Experiment 5.11.

For instance, given two electrified plastic straws A and B, we defined that the more electrified straw is the one which, at the same distance from the table or from the falling stream of liquid, attracts more pieces of paper or causes a greater curvature of the falling liquid (Figure 5.28). Body C in this case is a piece of paper or the water stream. The more electrified straw will also be that one which, at the same distance from an electric pendulum, causes a greater inclination of the pendulum from the vertical. The same definition can be applied to the other experiments described in this book.

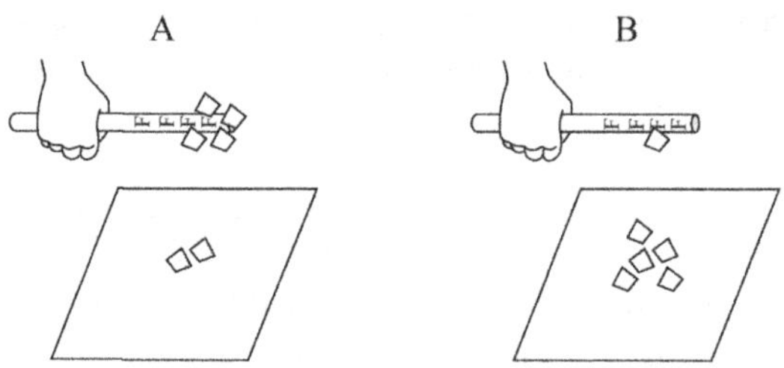

Figure 5.28: Straws A and B are at the same distance from a table. Straw A is more electrified than B because it attracts more pieces of paper.

In Section 5.6 we saw that the intensity of the force increases when the distance between the interacting bodies decreases. This suggests another procedure for measuring the electrification of a body. We define that the amount of charge of a body is indicated by the distance at which its electrical force creates visible effects. For instance, suppose we have two electrified plastic straws A and B of the same size. Let us suppose that A begins to attract pieces of paper at a distance of 15 cm from a table, while for B this attraction only begins when it is at a distance of 5 cm from the table. In this case we define that A is more electrified than B, that is, that A has a greater amount of electrical charge than B (Figure 5.29).

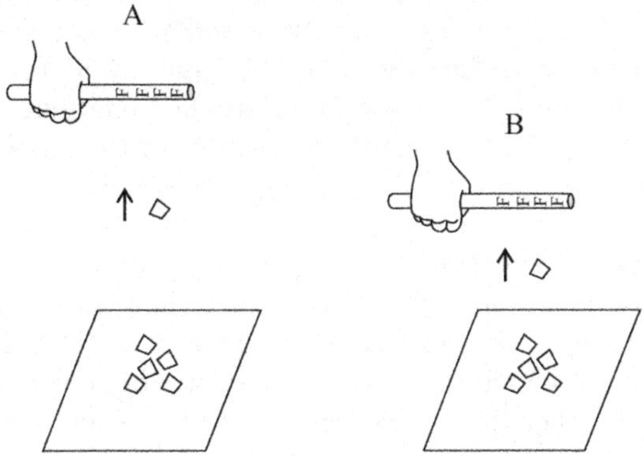

Figure 5.29: Straw A is more electrified than straw B because it begins to attract the pieces of paper at a greater distance from the table.

The other effects will also begin to be observed or detected at a greater distance for A than for B. For instance, let us consider that body A causes the falling water stream (or the electric pendulum) to curve when it is at a distance d_1 from the falling water (or the pendulum). For body B the equivalent distance will be d_2. When A is more electrified than B, then $d_1 > d_2$.

In conclusion, by definition, the electric force increases with the amount of charge.

We now present some simple but non trivial experiments.

Experiment 5.24

We choose two plastic straws electrified by friction. We place one of them horizontally, parallel to a table with many pieces of paper over the table. Let us suppose that it attracts N pieces of paper when it is at a distance d from the table. Let us suppose that the second electrified straw also attracts approximately N pieces of paper when it is at the same distance d from the table. By the previous definition of Section 5.7, we conclude that they have the same amount of charge.

We then move the straws away from the table. We join them side by side, horizontally. We move them together toward the table until they are at the same distance d from the table as before. We observe that together they attract a larger number of pieces of paper than each one of them separately.

When we join three or four equally electrified straws, the number of attracted pieces of paper increases even more at the same distance from the table.

Experiment 5.25

Let us imagine that we have two plastic straws which have been uniformly electrified by friction along their entire lengths. Let us suppose that they have approximately the same electrical charges, as indicated by the previous definition. We charge an electric pendulum by the *ACR* method by letting it touch one of the electrified straws. After this procedure, the pendulum begins to be repelled by the straw. We remove the straw and the electrified pendulum returns to the vertical. We place the straw horizontally at the same height as the disk of the pendulum. Let us suppose that the pendulum is inclined at an angle θ relative to the vertical when the closest tip of the straw is at a distance d from the vertical projection passing through the support of the pendulum, as in Figure 5.30 (a).

We join the two rubbed straws and bring them to the same distance d from the charged electric pendulum. We observe that the pendulum deviates by an angle greater than θ from the vertical, as in Figure 5.30 (b). This experimental fact, together with the previous definition, shows that two equally charged bodies, when placed together, have a larger charge than each one of these bodies individually.

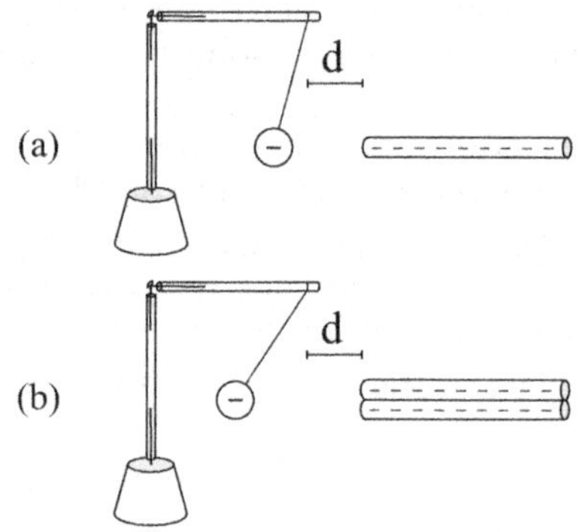

Figure 5.30: (a) The force upon a charged pendulum exerted by a charged straw is smaller than (b) the force exerted upon the same pendulum by two equally charged straws, at the same distance from the vertical passing through the point of support of the thread.

When we join three equally charged straws, we observe that the angle of inclination of the pendulum increases even more. This indicates a new increase in the force.

Definition: These observations allow a new definition. Suppose we have N bodies equally electrified with charges of the same sign. Suppose that we join these N charged bodies into a single set. We define the amount of charge of this set as N times the amount of charge of a single body.

Although this definition is very simple, it is not a trivial one. We illustrate this with an example. The level of a mercury thermometer can be defined as an indicator of the temperature of a body. For instance, we can define that two bodies A and B have the same temperature if the level of this thermometer connected to A is equal to the level of this thermometer connected to B. When the level of the thermometer is higher (lower) for A than B, then we define that A has a higher (lower) temperature than B. However, a simple experiment shows that when we join the two bodies A and B, in such a way that they touch one another, the level of the thermometer is not changed. Due to experiments like this we do not define the temperature of a set AB (that is, with A and B together touching one another) as twice the temperature of A.

As regards weights and electrical charges, many experiments show that when we bring together N bodies which have the same property, the gravitational and electrical effects produced by these N bodies will be greater than the effect produced by a single one of these bodies. Experiments like these lead to the previous definition. In other words, experiments like these make the previous definition a reasonable one.

Experiment 5.26

We tie the ends of two pieces of plastic straw having the same length and weight with a small silk thread. We can make two or three of sets of these. We rub each set in hair, but in some sets this friction is quicker and more intense. When we bring each of these sets near a negatively charged pendulum, we see that each one of them inclines the pendulum by a different angle, even when all of them are at the same distance from the vertical passing through the point of support of the pendulum's thread. This tells us that each system has different amounts of electric charge (Figure 5.31).

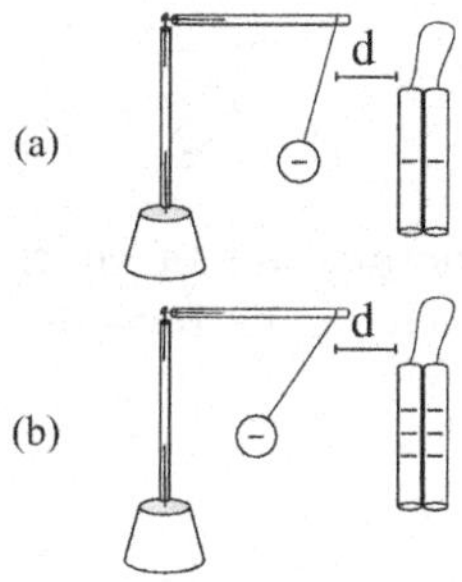

Figure 5.31: Bodies having different amounts of charge. The plastics in (b) have a greater electrical charge than the plastics in (a).

Each system can then hang from a horizontal support (like a pencil) by the silk thread. We observe that the straws do not remain vertical, but repel one another. Moreover, the system which repelled the charged pendulum with greater strength is also the system for which the two pieces of straw incline the most from the vertical (Figure 5.32). As the straws of all systems have the same weight, this shows once more that by increasing the charge they contain, we increase the electric force between them.

(a)

(b)

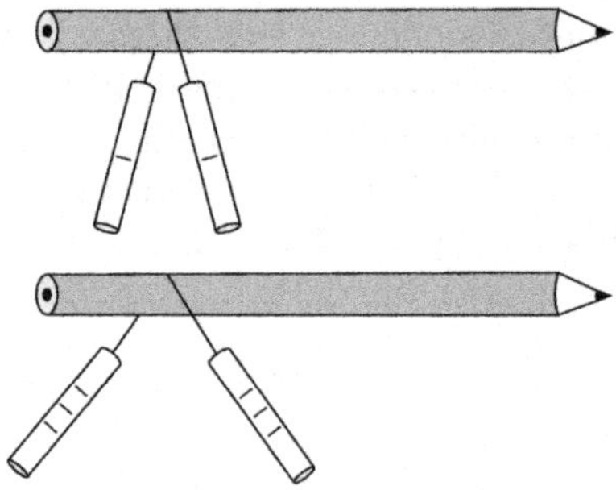

Figure 5.32: By increasing the amount of charge in each pair of straws, we increase the force between the charged straws of each pair.

Chapter 6

Conductors and Insulators

6.1 The Electroscope

We have already built a versorium and an electric pendulum. We now build another instrument, the *electroscope*. There are several varieties and we present only a few of them here. Sometimes the versorium and the electric pendulum are also called electroscopes, due to the fact that they indicate the presence of charged bodies in their neighborhood. But in this book we will reserve the name electroscope for the instrument described in this Section, and also discuss the properties that distinguish it from the earlier instruments.

Perhaps the simplest model is made of thin cardboard. We cut a rectangle with 7 by 10 cm sides, with the longer side vertical (Figures 6.1 and 6.2). We attach the rectangle to a plastic straw with two pieces of adhesive tape. The tape should be applied to the back side of the rectangle, not extending beyond the edges. The upper end of the straw should remain close to the upper edge of the rectangle, without extending beyond it.

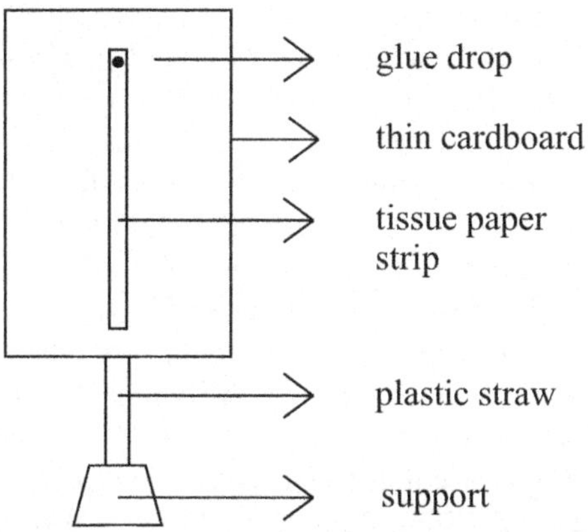

Figure 6.1: Electroscope seen face on.

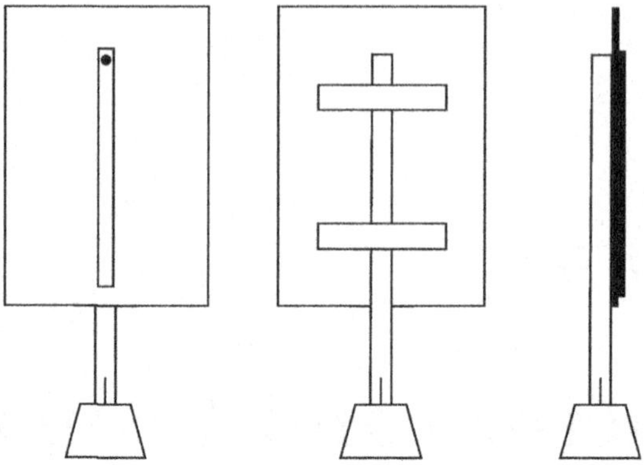

Figure 6.2: (a) Electroscope seen face on. (b) Back view. (c) Seen in profile.

We cut a very small strip of tissue paper, from 1 to 3 mm wide and 6 to 9 cm long. The effects to be described in this Chapter become more visible when utilizing a very thin and light strip. This tissue paper can be the kind used to build kites or employed to wrap fragile gifts. We glue the upper end of this strip to the upper middle of the rectangle. The strip should not be folded and should not go beyond the lower edge of the rectangle. Instead of glue, we can also attach it with a small piece of adhesive tape, provided the tape does not go beyond the edge of the rectangle.

We prepare a support for the electroscope such as the one in Section 4.4. *The crucial aspect of the electroscope is that it should have a plastic straw as the support for the cardboard.* That is, the cardboard should not be attached to, for instance, a wood or metal skewer. The plastic straw is then placed over the paper fastener of the support. If the rectangle does not remain in a vertical plane, we can put two straws together, one inside the other, in order to make it more rigid. In Section 6.5 we will present the fundamental components of an electroscope like this one, after performing several experiments with this one.

There are also several alternative electroscope models. We can cover the thin rectangular cardboard with aluminum foil and in place of the tissue paper we can also use aluminum foil. In addition, instead of gluing the tissue paper or attaching it with adhesive tape, we can also use a small metal hook attached at the rectangle. We then make a small hole with a needle in the upper portion of the strip and set it in the hook. The rectangles can also have other dimensions, like 2 by 8 cm, for instance. The dimensions of 7 by 10 cm utilized here are convenient for some of the experiments described in this work, as 7 cm is a little bigger than the diameter of the plastic coffee cup used as the support.

Some electroscopes have an upper cover, although this is not essential for the success of most experiments described here. Normally this cover is a disk of thin cardboard having a diameter equal to the side of the rectangle (in our example this would mean a disk with 7 cm diameter). We can also cover this upper disk with aluminum foil, but once more this is not essential. To attach the upper disk to the rectangle, there are several possibilities. (a) The simplest possibility is to fold the upper strip horizontally on the rectangle (0.5 or 1 cm wide and 7 cm long). The disk is then glued on this strip. (b) Another alternative way to attach the upper disk is to make a hole in its center. One leg of a paper fastener passes vertically through this hole and is attached inside the vertical plastic straw holding the rectangle. The second leg of the paper fastener should be placed horizontally around the disk, and attached inside a second hole made on an appropriate section of the disk.

The electroscope described in Figure 6.2 is composed of a single mobile strip, with the rectangle attached to the support. Another common model of electroscope has two mobile strips, or two mobile leaves. It is this model which is represented in most textbooks of electricity. The simplest way to make an electroscope like this is to glue together the upper ends of two strips of tissue paper at the lower edge of a rectangle. In Figure 6.3 we present an electroscope like this. Once more the thin cardboard rectangle is attached to a plastic straw with two pieces of adhesive tape on its back side. On the lower left side of the electroscope we glue the upper ends of two strips of tissue paper (or two thin strips of aluminum foil). They should be of the same length and their lower ends should be free to move away from one another. We can call this model a *classic electroscope.* If we wish we can also glue a few more pairs of strips along the lower edge of the rectangle.

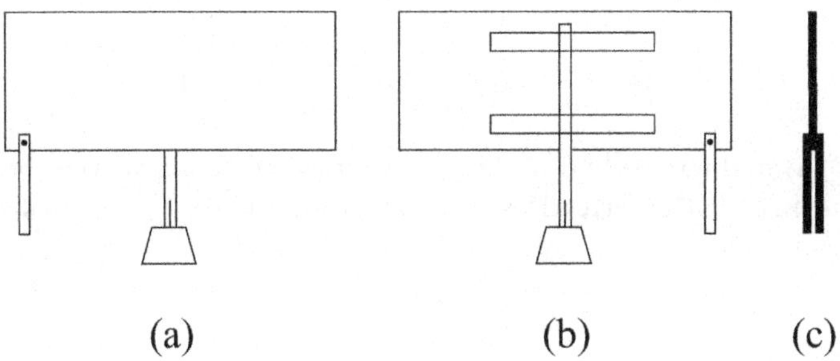

Figure 6.3: (a) Classic electroscope seen face on. (b) Back view. (c) Seen in profile, showing only the thin cardboard and the two strips.

We can also make a classic electroscope by folding a single tissue paper strip in the middle. We then support the strip through its center by a rigid horizontal support, like a metal wire. The two halves of the strip then hang then side by side, with the lower ends free to move away from one another. The metal wire should be supported from below by plastic straws, or it could be supported from above by silk threads.

The most refined model of the classic electroscope is made with two gold leaves. Normally it is surrounded by a glass cover to prevent perturbations due to air flow.

6.2 Experiments with the Electroscope

Experiment 6.1

We rub a plastic straw in hair and slowly bring it near the upper portion of the electroscope, without touching it. The strip moves away from the electroscope. When we remove the straw, the strip returns to its original vertical orientation along the thin cardboard (Figure 6.4).

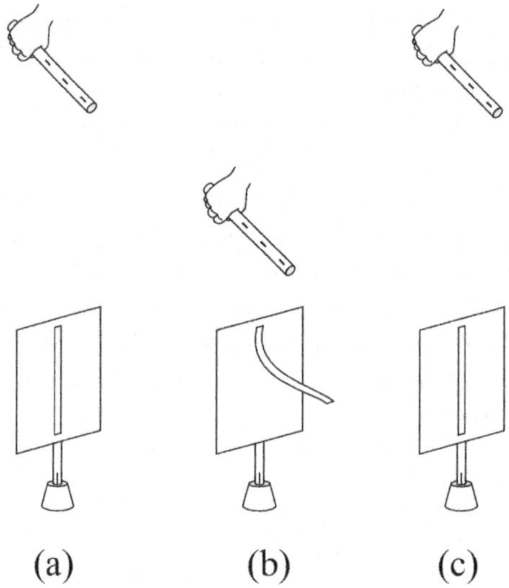

Figure 6.4: (a) and (b): When we bring a charged plastic near the electroscope, without touching it, the strip rises. (c) When we remove the electrified plastic, the strip drops down.

Experiment 6.2

We repeat Experiment 6.1, but we now scrape the rubbed straw on the upper edge of the electroscope. The ideal situation is to touch the thin cardboard with a portion of the rubbed straw which is close to the fingers holding it. We then scrape the straw while moving it away from the rectangle. The straw should be scraped toward the back side of the rectangle, and not toward the strip. This procedure can be repeated one or more times, also, if possible, spinning the straw while it is scraped. In order to facilitate the scraping, we can fix the electroscope by holding its vertical plastic straw. However, we should never touch the rectangle, which is made of thin cardboard, nor the paper strip. We observe that the strip moves away from the electroscope as it is rubbed.

We now move the rubbed straw away from the electroscope. In this case we observe that the strip remains raised, away from the rectangle (Figure 6.5).

Experiment 6.3

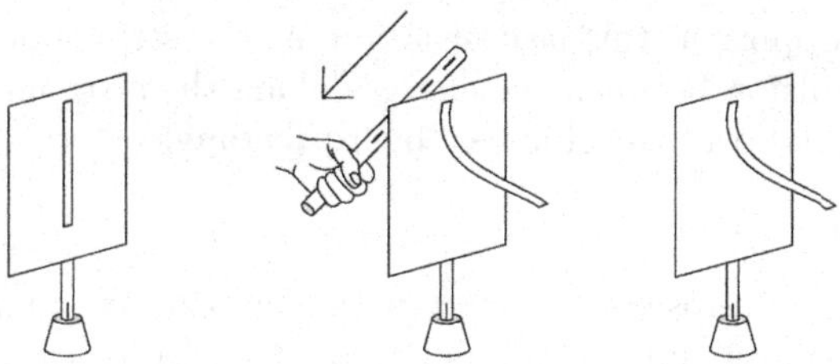

Figure 6.5: (a) Electroscope with its strip pointing downward. (b) We scrape the upper edge of the rectangle with a rubbed straw. (c) When the straw is removed, the strip remains away from the electroscope!

After we perform Experiment 6.2 and remove the rubbed straw, the strip remains away from the electroscope. When we slowly bring our finger laterally near the lower end of this strip, without allowing them to come into contact, we observe that the strip is attracted by the finger, moving toward it. When we remove the finger, we observe that the strip remains away from the electroscope.

From what we saw in Section 3.5, this means that the electroscope became electrically charged by the procedure of Experiment 6.2. It was charged by scraping a rubbed plastic over its edge. This experiment is analogous to Experiment 4.7.

But now we have something new. When we charged an electric pendulum by the *ACR* method, the pendulum was repelled by the rubbed plastic when we brought the plastic near the pendulum. However, when we removed the rubbed plastic, the pendulum thread returned to the vertical, although the pendulum remained charged. If we had not seen this experiment and looked at the vertical pendulum, we would not know whether the pendulum was charged or not. In order to test this condition, we might move his finger near the paper disk of the pendulum, without touching it. If the pendulum did not move, this would indicate that it was neutral. On the other hand, if the pendulum was oriented toward the finger, we would know that the pendulum had been charged by some mechanism. In order to know the sign of this charge, we might move another body with a known charge, positive for instance, near the pendulum. If a repulsion occurs, we will know that the pendulum is positively charged. If a strong attraction occurs, we will know that the pendulum is negatively charged.

The electroscope, on the other hand, presents a different behaviour. After it has been charged, its strip remains away from the rectangle even when the rubbed straw used to charge it has been moved far away from the system. This means that by simply looking at the strip of an electroscope, we can know if it is charged or not. When the strip points vertically downward, together with the rectangle, the electroscope is discharged. When the strip is raised, away from the rectangle, we know that the electroscope has been charged. What prevents the strip from dropping in this last situation, as the strip should do due to its weight, is the repulsion between the charges along the strip and the charges of the same sign spread over the thin cardboard rectangle.

Experiment 6.4

We charge the electroscope as in Experiment 6.2. We remove the rubbed plastic and the strip remains raised, away from the electroscope. We now move our finger near the upper edge of the electroscope and touch it. Immediately the strip drops, sticking to the rectangle. When we remove the finger, the strip remains vertical (Figure 6.6).

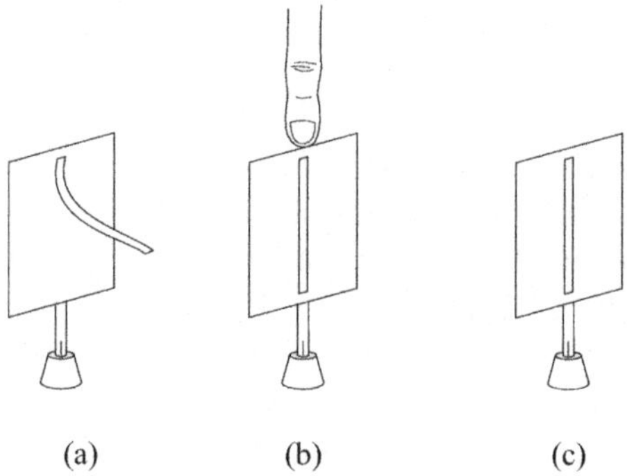

Figure 6.6: Discharging an electroscope by touching it. (a) An initially charged electroscope. (b) When we touch the cardboard's upper edge, the strip drops. (b) The strip remains vertical after we remove the finger.

When we again bring a finger near the lower end of the strip from the side, we observe that the strip no longer moves toward the finger, as it had done before in Experiment 6.3. We conclude that we have discharged the electroscope when we touch the thin cardboard, as was the case in Experiment 4.9. This is due to the electrical grounding, Section 4.5.

Experiment 6.5

We charge the electroscope by scraping it with a rubbed straw, as in Experiment 6.2. We again use the rubbed straw we used earlier to charge the electroscope. In particular, we slowly bring the straw near the raised strip of the electroscope. The rubbed straw should be horizontal, at the same height of the lower end of the raised strip. The motion of approach should be very slow, in order to prevent them from coming into contact. We should attentively observe the direction in which the strip tries to move, that is, whether it moves toward the rubbed strip or away from it. By performing this experiment carefully, we observe that the strip moves toward the rectangle, that is, it moves away from the approaching rubbed straw.

We can even cause the strip to move to and fro with the rubbed straw, moving it toward the strip and away from it. The strip will move at the same speed, toward the electroscope and away from it. If we are to observe this oscillatory motion of the strip, the amplitude motion of the rubbed straw should be low. That is, we should use small movements, and avoid bringing it very close to the strip (Figure 6.7).

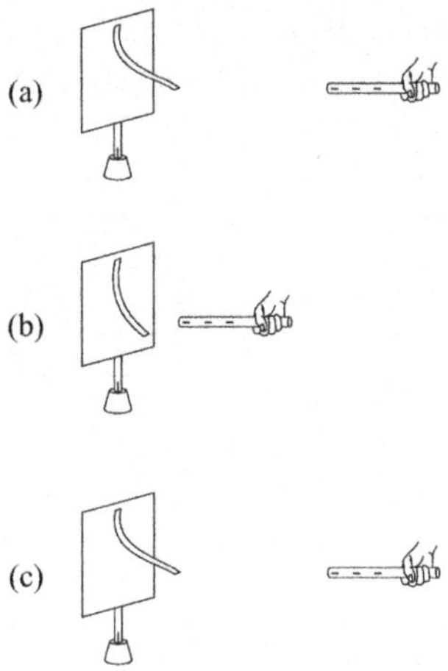

Figure 6.7: Repulsion between the charged plastic and the electroscope charged by this rubbed plastic. (b) When the plastic is moved near the electroscope, the strip drops. When the plastic is moved away from the electroscope, the strip raises, (a) and (c).

This shows that in Experiment 6.2 the electroscope has become electrified with a charge of the same sign as that of the rubbed straw, as there is a repulsion between them. We can then represent the charges of the electrified electroscope as having the same sign as the charges of the rubbed plastic which was scraped on the thin cardboard (Figure 6.8).

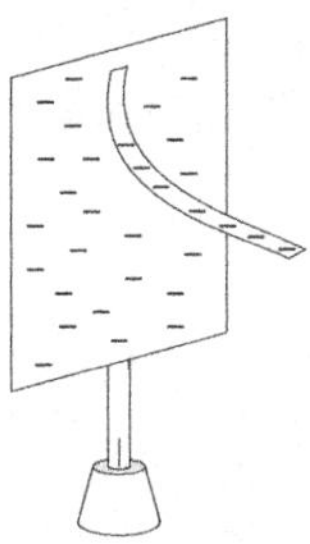

Figure 6.8: Charges spread over an electroscope which has been electrified with a negative straw.

Experiment 6.6

We repeat Experiment 6.2 charging an electroscope negatively by scraping it with a straw rubbed with hair. When we remove the straw, the strip remains raised. We charge another straw positively, as in Experiment 5.15. We slowly bring this second straw near the electroscope, preventing them from coming into contact. In this case there is an attraction between them. This attraction is so strong that we can even cause the strip to rise up past the upper edge of the electroscope, making it follow the positive straw (Figure 6.9)!

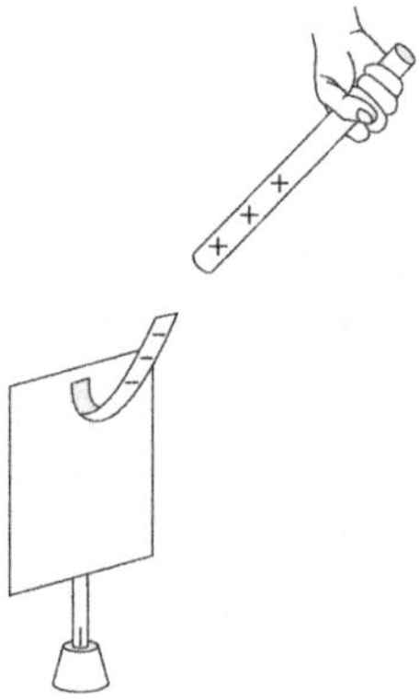

Figure 6.9: Attraction between a positive straw and a negative electroscope.

Experiment 6.7

We charge two electroscopes by scraping them with straws rubbed with hair, as in Experiment 6.2. We remove the straws and the strips remain raised. We place these two charged electroscopes facing each other, in parallel planes, with both strips raised toward one another. We slowly bring the electroscopes near one another, without allowing the strips to come into contact. We observe that they tend to move away from one another, each one of them returning back to the rectangle of its own electroscope.

This once more shows the repulsion between charges of the same sign.

Experiment 6.8

We charge an electroscope negatively by scraping it with a straw rubbed with hair, as in Experiment 6.2. Another electroscope is charged positively by scraping it with another straw rubbed between two hard rubber hoses. We remove the straws and the strips remain raised. These two electroscopes are placed in parallel and facing each other, with the strips pointing toward one another. We slowly move both electroscopes close to one another, preventing the strips from coming into contact. This time we observe that the strips attract one another, with each strip moving away from its own cardboard (Figure 6.10).

Experiment 6.9

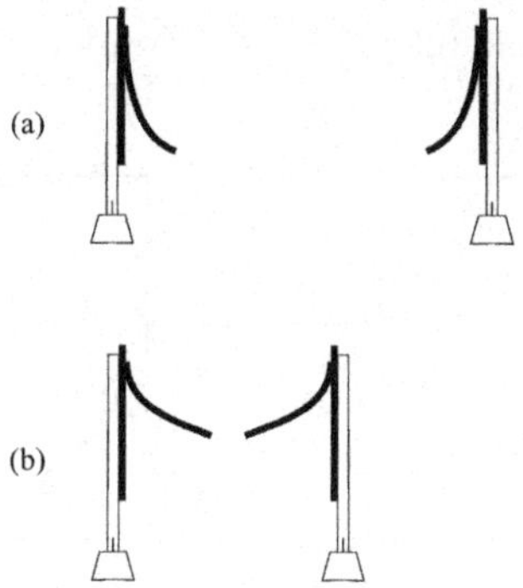

Figure 6.10: Attraction between the strips of two oppositely charged electroscopes.

We begin the experiment with an initially discharged electroscope. We rub a straw in hair. A small portion of the rubbed straw touches the upper edge of the electroscopes and scrapes against it, as in Experiment 6.2. We observe that the strip rises by an angle θ_1 away from the electroscope.

Without touching the thin cardboard of the electroscope or the strip of paper, we again scrape the rubbed straw on the upper edge of the electroscope. When we remove the straw the strip remains raised, but now inclined by an angle θ_2 greater than θ_1. That is, $\theta_2 > \theta_1$. This is shown in Figure 6.11.

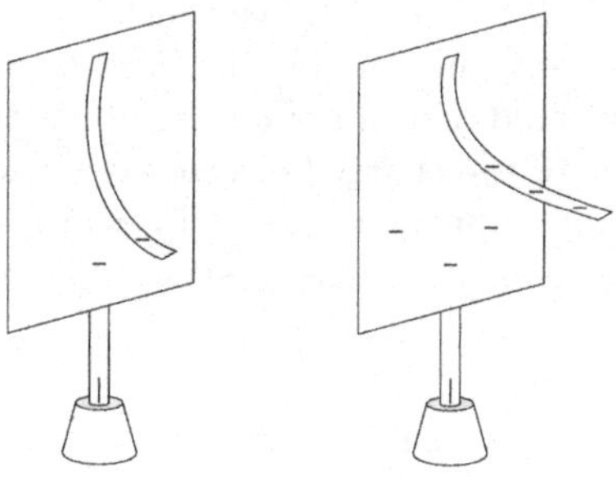

(a) (b)

Figure 6.11: The angle of displacement of a strip in relation to the electroscope is a measure of the amount of charge on it.

This procedure can be repeated a few more times. During this procedure the straw should be rubbed one or more times in hair.

This experiment indicates that the electroscope can be used to measure the amount of charge, according to the definitions of Section 5.7. That is, by increasing the amount of charge on it, we increase the angle between the strip and the rectangle.

Another way of illustrating this effect can be seen with the classic electroscope. By increasing its amount of charge, we increase the angle of separation of its two leaves (Figure 6.12).

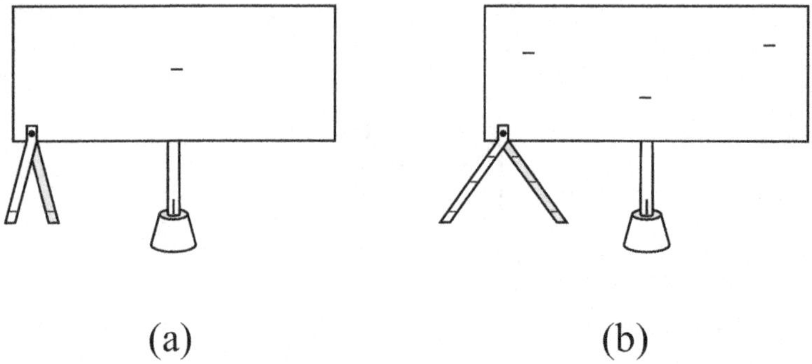

Figure 6.12: The angle between the two leaves of an electroscope is a measure of the amount of charge on it.

6.3 Which Bodies Discharge an Electroscope by Contact?

Experiment 6.10

We saw in Experiment 6.4 that when we touch a charged electroscope with our finger, the electroscope is immediately discharged. We charge it once more as in Experiment 6.2. We hold one end of a neutral plastic straw with our hand and touch the upper edge of the charged electroscope with the other end of the straw. In this case nothing happens to the strip, which remains raised (Figure 6.13). We conclude that a neutral plastic straw does not remove the charge from the electroscope.

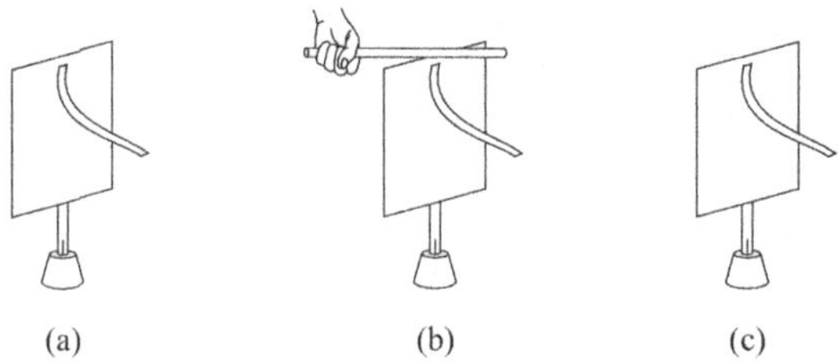

Figure 6.13: (a) An initially charged electroscope. (b) We hold one end of a neutral plastic straw with our hand and touch the edge of the electroscope with the other end of the straw. Nothing happens to the strip. (c) When we remove the straw, the strip remains raised. That is, an electrified electroscope is not discharged when we touch it with a neutral plastic straw in our hand.

Experiment 6.11

We again charge the electroscope, as in Experiment 6.2. This time we hold one end of a wood skewer in our hand. We touch the upper edge of the charged electroscope with the other end of the wood skewer. We observe that the strip drops immediately (Figure 6.14). When we remove the wood skewer, the strips remains vertical. We conclude that the wood skewer has removed the charge on the electroscope.

Figure 6.14: (a) An initially charged electroscope. (b) We hold one end of a wood skewer with our hand and touch the upper edge of the electroscope with the other end of the skewer. The strip drops immediately. (c) When we remove the skewer, the strip remains down. That is, an electrified electroscope is discharged when we touch it with a piece of wood in our hand.

6.3.1 Definitions of Conductors and Insulators

Experiments 6.10 and 6.11 (Figures 6.13 and 6.14) present a fundamental distinction between substances. Due to the importance of these properties, two names were created in order to classify substances into two groups.

Definitions: Substances which discharge an electrified electroscope simply by touching it are called *conductors*. Substances which do not discharge the electroscope are called *insulators*, *nonconductors*, or *dielectrics*.

The discovery of these two kinds of substances is due to Gray in 1729. He also discovered some of the main properties of these substances, publishing his results in a fundamental paper in 1731, which we will discuss in greater detail in Appendix B.[1] The expressions *conductor* and *insulator* appear to be due to Jean Théophile Desaguliers (1683-1744).[2] One of Desaguliers's quotations mentions this:[3]

> In the following account, which is the sequel of former experiments, I call *conductors* those strings, to one end of which the rubb'd [glass] tube is applied; and *supporters* such horizontal bodies as the *conductor* rests upon.

[1] [Grah].

[2] [Desa] (referring to [Desc]), [Pri66, p. 82], and [Hei99, pp. 292-293, note 12].

[3] [Desa, p. 193].

Before Desaguliers, Du Fay had already used the expression *insulated* in order to refer to a conductor supported by bodies which do not allow electricity to escape through them. In 1733 Du Fay discussed the transmission of electricity along strings suspended by silk threads, a phenomenon which had been discovered by Gray. In this work Du Fay said the following, our emphasis in italics:[4]

> This experiment proves how necessary it is that the [conducting] string utilized to transmit far away the electricity, be *insulated*, that is, [the conducting string should be] suspended only by bodies which are the least possible appropriate to charge themselves of electricity.

In 1737 he said the following, our emphasis in italics:[5]

> Therefore, being sure of this equality for the experiments which I intended to make, I utilized an iron bar, one square inch in section [2.54 by 2.54 cm], and four feet in length [122 cm]. It was, as I said, suspended upon silk cords and *insulated*, in order that nothing could deviate the electric vortex which would be communicated to it [by the rubbed glass tube].

It is possible that the expression *insulator* had its origins in these quotations from Du Fay. In Appendix B we discuss Gray's work in more detail.

Experiment 6.12

We now repeat Experiments 6.10 and 6.11 in order to discover which substances are conductors and which are insulators. The procedure will always be the same. We charge an electroscope by scraping it with a rubbed plastic straw, as in Experiment 6.2. We remove the straw. We then hold a certain object with our hands and touch the upper edge of the thin cardboard rectangle with some part of this object. If the electroscope discharges, we say that the substance is a conductor. If the electroscope does not discharge, we say that the substance is an insulator.

This experiment can be done with single threads of several materials: cotton, silk, polyester, synthetic polyamide, hair, copper, etc. It can also be done with solid materials like metal, wood, glass, rubber, plastic, paper, tissue paper, etc.

In some cases it is easier to hold the supporting vertical plastic straw of the charged electroscope, and then touch the corner of the thin cardboard to the test substance, like a wall, blackboard, piece of furniture, etc. As always, we must avoid touching the thin cardboard rectangle and in its strip with our hands, to prevent the electrical grounding through our body.

---

[4][DF33d, p. 249]: Cette expérience prouve combien il est nécessaire que la corde dont on se sert pour transmettre au loin l'électricité, soit isolée, ou ne soit soûtenuë que de corps le moins propres qu'il est possible à se charger eux-mêmes de l'électricité.

[5][DF37b, p. 94]. M'étant donc assüré de cette égalité pour les expériences que j'avois dessein de tenter, je me suis servi d'une barre de fer d'un pouce en quarré & de quatre pieds de long, elle étoit, comme je l'ai dit, suspenduë sur des cordons de soye & isolée, afin que rien ne pût détourner le tourbillon électrique qui lui seroit communiqué par le tube.

Experiment 6.13

The procedure described in the final paragraph of Experiment 6.12 is also appropriate to determine which liquids are conductors or insulators. Before beginning these tests, we take an empty receptacle which will later be filled with different liquids. We need to use a conducting receptacle. In order to know if the receptacle is a conductor or an insulator, we charge an electroscope and touch its thin cardboard against the receptacle. If the electroscope discharges after this contact, this means that the receptacle is really conducting. Examples of suitable receptacles are ones made of metal, glass, or wood. We can then continue with the experiment.

Initially we completely fill the receptacle with the liquid to be tested. We now illustrate what happens with a conducting liquid like water (Figure 6.15). In (a) we have an electrified electroscope. In (b) we submerge an edge of the electroscope in the water. The cardboard should not touch the receptacle, only the water. We observe that the strip drops. (c) When we remove the electroscope from the water, the strip remains down.

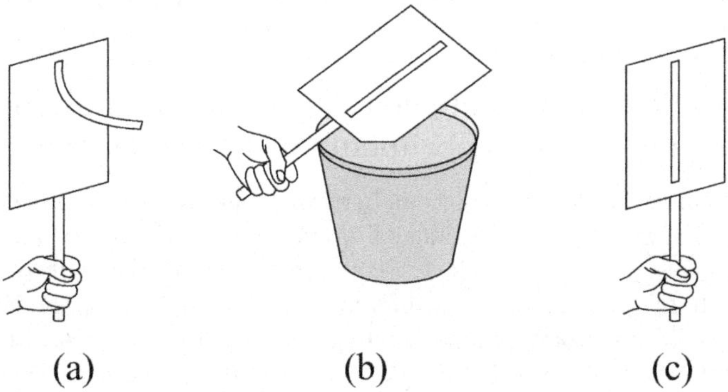

Figure 6.15: (a) An initially charged electroscope. (b) We submerge an edge of the electroscope in the water. Its strip drops. (c) When we remove the electroscope from the water, the strip remains down.

In Figure 6.16 we illustrate what happens in the case of an insulating liquid like kitchen vegetable oil. In (a) we have an electrified electroscope. We hold it only by its plastic straw to avoid touching the thin cardboard or the raised strip. We then submerge an edge of the cardboard in the liquid. The cardboard should not touch the receptacle, only the oil. We observe that the strip remains raised, as in (b). When we remove the electroscope from the liquid, its strip remains raised, as in (c).

The same procedure can also be applied to determine whether other substances, such as flour or cornmeal, are conducting. That is, a conducting receptacle is filled with flour. We submerge the edge of the charged electroscope in the flour and observe the behaviour of its strip. During this procedure it is important to avoid touching the thin cardboard rectangle or its strip with the

(a) (b) (c)

Figure 6.16: (a) An initially electrified electroscope. (b) We submerge an edge of the cardboard in an insulating liquid. The strip remains raised. (c) When we remove the electroscope from the liquid, the strip remains raised.

conducting receptacle and with our hand in order to prevent discharge of the electroscope.

6.3.2 Bodies which Behave as Conductors and Insulators in the Usual Experiments of Electrostatics

In order to make a clear distinction between conductors and insulators, the ideal procedure would be to consider all bodies of the same shape and size. For instance, we could touch the electroscope with several cylinders of the same length and diameter, but made of different substances. But for the time being, however, we will not worry about the shape or size of the bodies to be tested.

The final result of Experiments 6.10, 6.11, 6.12, and 6.13 performed with many substances is as follows:

- **Conductors based on the usual experiments of electrostatics:** Humid air, human body, all metals, paper, thin cardboard, tissue paper, aluminum foil, wood, a piece of chalk, most glasses at ambient temperature, porcelain, fresh water, alcohol, shampoo, kerosene, milk, soft drinks, detergent, wall, blackboard, cork, leather, wheat flour, corn flour, acrylic thread, salt, sugar, sawdust, earth or clay, brick, most kinds of rubber, etc.

- **Insulators based on the usual experiments of electrostatics:** Dry air, amber, plastic, PVC, silk, heated glass, nylon or synthetic polyamide, polyester, wool, a single human hair, acrylic tube, Styrofoam, a chocolate bar, cooking vegetable oil, ground coffee, and a few kinds of rubber.

The number of conducting substances is much larger than the number of insulating substances. From these two lists we can see that most substances are conductors, very few are insulators. Some of the conductors are very good, discharging almost instantaneously the electroscope, as is the case of the human body, metals, cotton, or paper. Although wood is a conductor, it does not conduct as well as the human body. This is indicated by the longer time required to discharge the electroscope when we touch it with a wood, compared with the very short time in which it is discharged when we touch the electroscope with our body or with a piece of metal.

Glass must be considered separately. The majority of the common glasses discharges the electroscope, although more slowly than the metals. On the other hand, when heated in fire or in a microwave, they can behave as insulators. In other words, after being heated they do not normally discharge the electroscope, or they discharge it much more slowly than at normal temperature. Normally they behave as conductors due to the humidity or water vapor accumulated over their surface. When the glass is heated, this water is evaporated and they then behave as insulators. Many old scientists, like Gray and Du Fay, used rubbed glass tubes in their experiments, holding them by their hands. The tubes behaved as an insulator. They often mention the need to heat the tubes before rubbing, as this heating procedure increased the amount of charge they acquired. It also caused the acquired charge to last longer on their surface. The other reason why their glasses could behave as an insulator was probably due to their chemical composition. Normally they utilized flint-glass, which is a kind of glass that has lead in its composition. This kind of glass is much more insulator than the majority of glasses found nowadays in stores. It is not easy to find flint-glasses nowadays, except in specialized stores.

On dry and cold days the electroscope can remain charged for several minutes. In this case the air around the electroscope behaves as a good insulator. On hot and humid days, on the other hand, and especially on rainy days, it is difficult to keep the electroscope charged. It normally discharges as soon as it is electrified by being scraped with a rubbed straw. The air around it behaves as a conductor in these situations. For this reason most experiments work well on dry days, when the charges generated in insulators can be maintained for longer times. Also the charges located in conductors electrically insulated from the ground can be stored for a longer time in dry weather, as compared with the short time they can be stored in humid weather. On humid days many experiments do not work as expected, or the effects are not so visible or detectable.

Most kinds of rubber behave as conductors. This can be due to the humidity over their surfaces, to their chemical composition, or to their fabrication processes. A few kinds of rubber behave as insulators. For this reason the ideal procedure is to test all available substances. Only after a test like the one described here can we classify them depending on their behaviour.

Dividing the substances into conductors and insulators is one of the most important aspects of the whole science of electricity. Together with the existence of positive and negative charges, with attractions and repulsions, this fact allows us to understand a whole set of phenomena.

6.4 Which Bodies Charge an Electroscope by Contact?

Experiment 6.14

Two initially discharged electroscopes are placed in parallel, facing one another, with the strips on the outside. There should be a distance of approximately 15 cm between them. A neutral plastic straw is placed above the thin cardboards of the electroscopes, supported by them, as in Figure 6.17. After this preparation, we take a second plastic straw, electrifying it by friction in our hair.

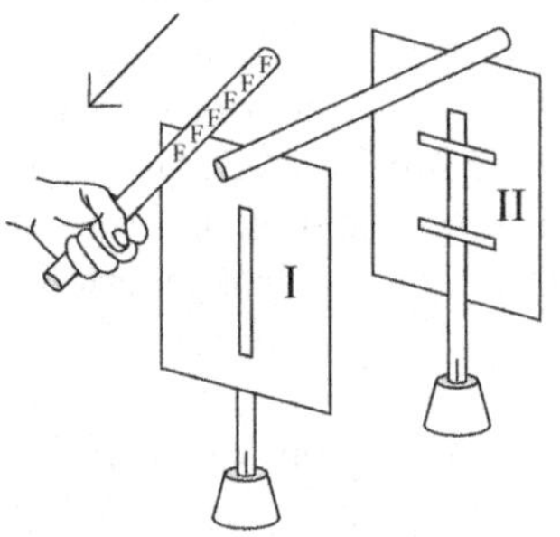

Figure 6.17: Two initially discharged electroscopes with their strips on the outside. They are connected by a neutral straw. A second charged straw is scraped against the cardboard of electroscope I.

This second rubbed straw is scraped against the cardboard of electroscope I of Figure 6.17. We observe that only the strip of electroscope I rises, as the strip of electroscope II does not move. This sequence of procedures in illustrated in Figure 6.18.

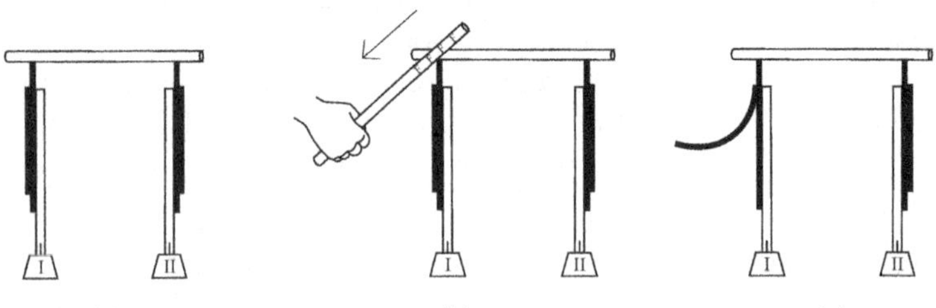

Figure 6.18: (a) Two initially discharged electroscopes connected by a neutral straw. (b) A second electrified straw is scraped against the upper edge of electroscope I. (c) The charged straw is removed. We observe that only the strip of electroscope I rises.

Experiment 6.15

The procedure of Experiment 6.14 is repeated. But now the two initially neutral electroscopes are connected by a wood barbecue skewer (Figure 6.19).

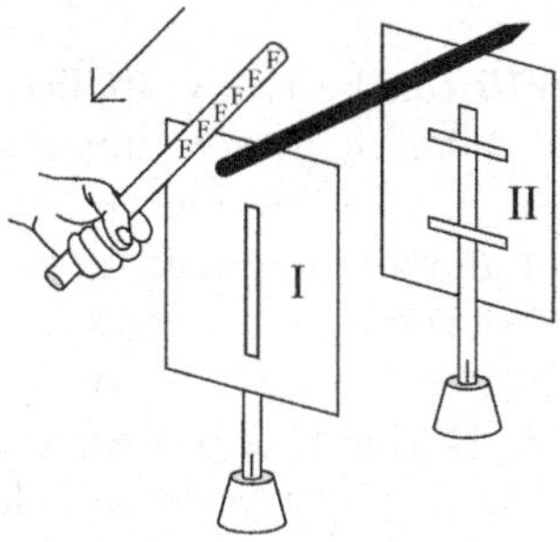

Figure 6.19: Two initially discharged electroscopes are connected by a wood skewer. An electrified straw is scraped only against electroscope I.

An electrified straw is scraped only against electroscope I of Figure 6.19. In this case we observe that both strips rise. This sequence of procedures is illustrated in Figure 6.20.

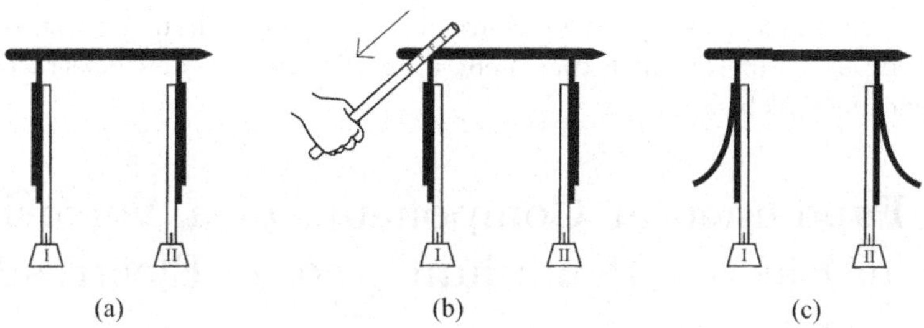

Figure 6.20: (a) Two initially discharged electroscopes connected by a wood skewer. (b) We scrape only electroscope I with an electrified straw. (c) The electrified straw is removed. We observe that both electroscopes have been charged.

Experiments 6.14 and 6.15 are the opposite of Experiments 6.4, 6.10, 6.11, and 6.12. In these cases we saw which bodies did or did not discharge an electrified electroscope, by touching this electroscope with this substance in our hand. We are now analyzing which substances do or do not charge the initially discharged electroscope II, when we connect it through this substance to electroscope I which is charged by scraping it with an electrified straw. The result is the same. That is, a neutral plastic straw does not discharge an electrified electroscope when touched against it, as in Experiment 6.10. It also does not charge electroscope II when this body is the material link between electroscopes I and

II, when electroscope *I* is charged by being scraped with an electrified straw, as in Experiment 6.14. On the other hand, a wood skewer charges electroscope *II* in this case, as in Experiment 6.15.

Experiment 6.16

Experiments 6.14 and 6.15 can be easily applied to other substances. We can, for instance, connect them through their upper edges with a wire or thread (of cotton, polyester, copper, etc.), by a stick (of wood, metal, plastic, PVC, etc.), by a strip (of paper, of aluminum foil, of tissue paper, of cloth, etc.), and so on. After this procedure, we rub a plastic straw. We then utilize this rubbed plastic to charge the first electroscope by scraping it with the rubbed straw, as described in Experiment 6.2. During this process we observe the behaviour of the second electroscope. If the strip of this second electroscope does not move, and remains pointing downward, this means that the substance connecting the two electroscopes did not allow the transfer of charges between them. That is, this substance is an insulator. On the other hand, if the strip of the second electroscope moves away from it, and remains raised, this means that there was a transfer of charges between the two electroscopes. Part of the charge accumulated in the first electroscope was then conducted through the connecting substance to the second electroscope. By performing this experiment, we observe that the substances considered as insulators in Experiment 6.12 do not allow the second electroscope to be charged. On the other hand, the substances considered as conductors in Experiment 6.12 allow the second electroscope in this experiment to be charged.

6.5 Fundamental Components of a Versorium, an Electric Pendulum, and an Electroscope

Now that we know the distinction between conductors and insulators, together with their main properties, we can understand the structure of the instruments constructed so far.

In the case of a metal versorium we have an horizontal conducting needle (like the steel paper fastener) (Figures 3.4 and 6.21 (a)). Normally it is supported on a vertical conducting pin attached to a wooden board or cork. That is, all elements of this versorium are conductors. The plastic versorium has an insulating hat (Figures 3.5 and 6.21 (b)). The Du Fay versorium, on the other hand, is composed of an insulating plastic hat, and has a conductor at one of its tips, the aluminum foil (Figures 4.22 and 6.21 (c)).

The electric pendulum is composed of an insulating silk thread with a conductor at its lower end, the paper disk (Figure 6.22). The silk thread is crucial. It is this thread, together with the plastic straws that make up the pendulum, which prevent the charge acquired by the *ACR* method from discharging to the ground. If it were not for this silk thread and plastic straws, the paper disk could not remain charged after contact with a rubbed material. If the paper

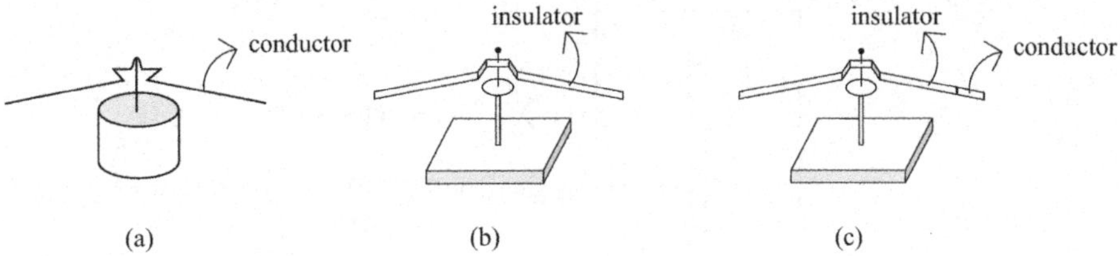

Figure 6.21: Components of a versorium. (a) Metal versorium. (b) Plastic versorium. (c) Du Fay versorium.

disk were tied, for instance, in a cotton thread supported by a wood skewer, we would not be able to see the mechanism of attraction, contact, and repulsion.

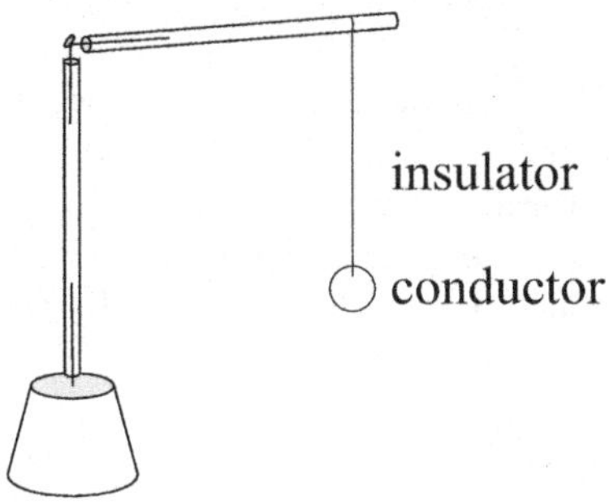

Figure 6.22: Components of an electric pendulum.

Nylon (synthetic polyamide) and polyester threads also behave as insultors. Therefore we can use these threads to make an electric pendulum. But we should not use a cotton thread.

The electroscope, on the other hand, is composed of two conductors, the thin cardboard rectangle and the tissue paper strip, supported by an insulator, the plastic straw (Figure 6.23). This plastic is crucial. It is the insulating property of the plastic which prevents the discharge of an electrified electroscope to the ground.

If we had a wood skewer instead of a piece of plastic, the electroscope would always discharge to the ground after being scratched with a rubbed straw. Therefore, it would not be possible to keep it electrified after being charged.

6.6 Influence of the Electric Potential Difference upon the Conducting or Insulating Behaviour of a Body

In Experiments 6.4, 6.10, and 6.12 the materials were classified as conductors (insulators) if they discharged (did not discharge) an electrified electroscope.

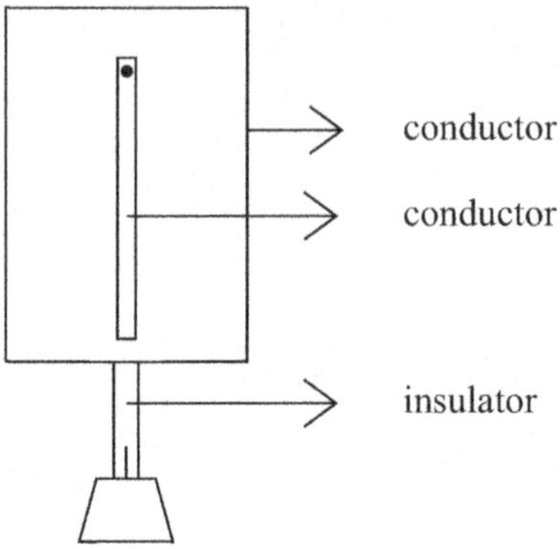

Figure 6.23: Components of an electroscope.

Another way to make this classification is to observe whether or not the materials discharge an electric battery.

Definitions: The materials which discharge a battery through a connection of the positive terminal of the battery with its negative terminal are called *conductors*. The materials which do not discharge the battery are called *insulators*. Let us see how to make this classification.

Experiment 6.17

Materials used in this experiment are represented in Figure 6.24. We utilize three pieces of insulated copper wire, uninsulated at their ends (Figure 6.24 (a)). A new large alkaline battery, D size, which generates a potential difference of 1.5 volts between its poles (Figure 6.24 (b)). It is also helpful to employ a battery support, in order to facilitate its electrical connections with the wires. We also utilize a small 1.5 volt bulb and socket. A switch is also helpful, although this is not essential (Figure 6.24 (c)). These items can be found in stores devoted to electricity and electronics.

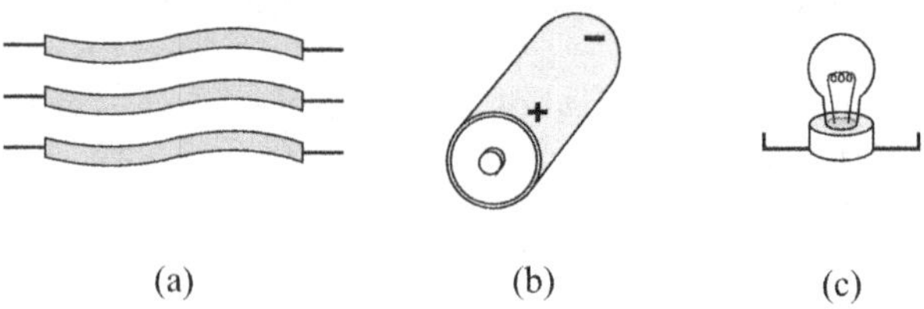

Figure 6.24: (a) Three pieces of insulated copper wire (strip the ends). (b) A new D size battery. (c) A 1.5 volt bulb and socket.

One uninsulated end of the first wire is connected to the negative terminal of the battery. The other end of this first wire will be shaped in a hook in Figure 6.25. This free end of the first wire will be called A. One uninsulated end of the second wire is connected to the positive terminal of the battery, with the other end connected to one of the terminals of the socket. One end of the third wire is connected to the other terminal of the socket. The other end of this third wire will make another hook. We will call this free end of the third wire B. The distance between A and B should be around 10 cm (Figure 6.25).

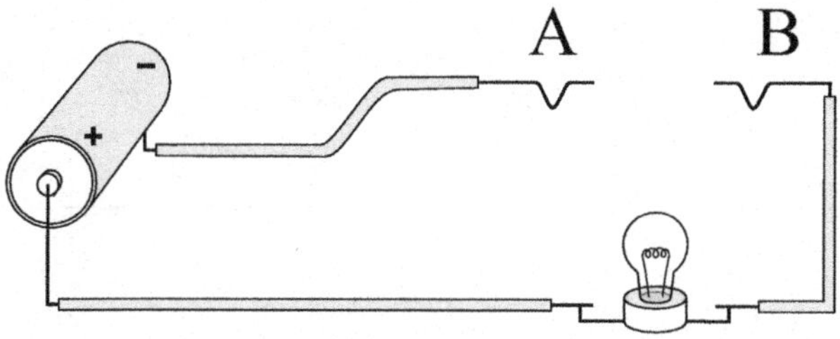

Figure 6.25: Circuit tester.

When everything is ready, we take a fourth piece of copper wire, uninsulated at the ends. One end of this fourth wire is connected to A and another end to B. The bulb should turn on. This will indicate that the electrical connections or contacts are properly made. Moreover, this will indicate that there is an electric current through the wires and bulb, as illustrated in Figure 6.26.

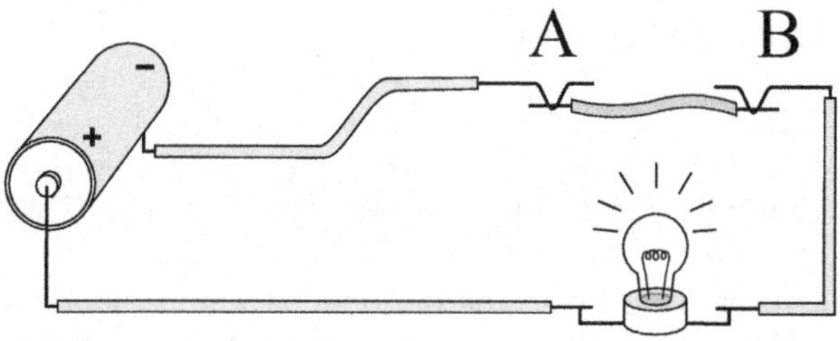

Figure 6.26: When the uninsulated ends of a copper wire are connected to ends A and B, the bulb turns on.

Since the bulb turns on, it is said that the copper wire is a conductor of electricity when it is under a potential difference of 1.5 volts = 1.5 V.

If the bulb is kept turned on for several minutes, the battery gets weaker. This is indicated by the intensity of the light, which decreases and goes to zero. When this happens the battery has been discharged. To prevent it from discharging, the best option is to open the contact (that is, remove the fourth wire) as soon as the bulb light turns on.

Experiment 6.18

Before conducting this next procedure, it is essential to ensure that Experiment 6.17 works properly with each of the materials we will be using. This will indicate that all electrical contacts are well made and working properly. We will suppose that this is the case. We will also assume that that the battery is still new and charged after the fourth wire has been removed.

We now connect points A and B with a plastic straw. When we do this, the light bulb does not turn on. This indicates that no electric current is flowing through the circuit (Figure 6.27 (a)).

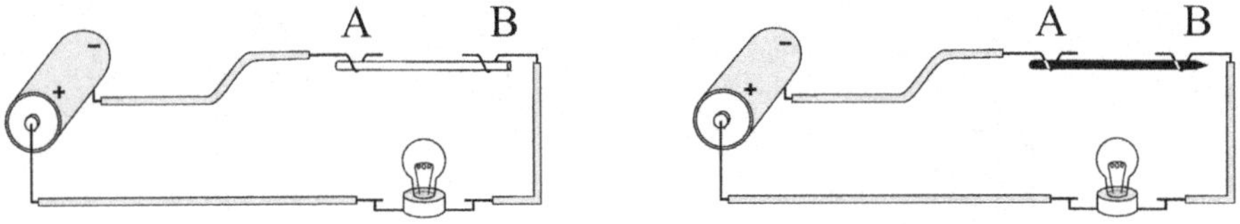

(a) (b)

Figure 6.27: (a) The light bulb does not turn on when we connect A and B with a plastic straw. (b) The bulb also stays off when A and B are connected with a wood skewer.

Definition: plastic is an insulating material when it is under a potential difference of 1.5 volts. The reason for this definition is that it does not allow the discharge of the battery when it connects its positive and negative terminals. This is indicated by the fact that the battery does not turn on when A and B are connected with a plastic straw.

Experiment 6.19

We repeat Experiment 6.17 but we now connect A and B with other substances.

For convenience we could place an optional switch in the middle of the first or second wire. With this switch we can open or close the electrical circuit at will. We open the switch and place the substance to be tested between A and B. We then close the switch and observe the bulb.

Definitions: When the light turns on, the substance is called a conductor. When the light does not turn on, the substance is called an insulator. We can test all substances listed in Experiment 6.12.

This is illustrated in Figure 6.27 (b) where A and B are connected with a wood skewer. In this case the light bulb does not turn on. According to our definition, this indicates that wood is an insulator when it is under a potential difference of 1.5 volts even though the wood skewer is a conductor at much higher voltages.

Experiment 6.20

The best way to test liquids is to obtain an insulating receptacle (like a plastic cup, for instance). Initially it should be empty. In order to verify if it really is an insulator, we connect A and B with this empty receptacle. We will suppose that the light does not turn on, indicating that it is an insulator at 1.5 volts.

The plastic cup is then filled with the liquid to be tested. After this, terminals A and B are submerged in the liquid. In Figure 6.28 we illustrate what happens with fresh water from a tap or rainwater.

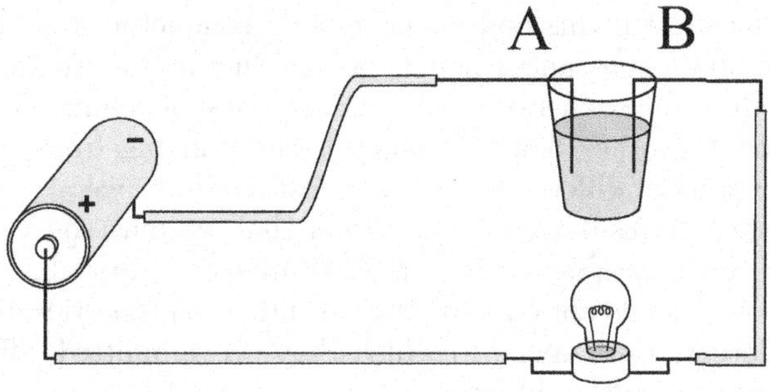

Figure 6.28: The light bulb does not turn on when A and B are connected with fresh water.

That is, the light does not turn on, indicating that fresh water is an insulator when it is under a potential difference of 1.5 volts.

6.6.1 Substances which Behave as Conductors and Insulators for Small Potential Differences

We perform several tests analogous to Experiments 6.17 to 6.20. The final result is as follows:

- **Substances which behave as conductors when under a potential difference of 1.5 V:**
  All metals.

- **Substances which behave as insulators when under a potential difference of 1.5 V:**
  Dry air, humid air, amber, plastic, silk, wood, heated glass, glass at room temperature, nylon or synthetic polyamide, PVC, polyester, wool, human hair, acrylic tube, acrylic cloth, Styrofoam, a chocolate bar, ground coffee, paper, thin cardboard, tissue paper, a piece of chalk, porcelain, fresh water, alcohol, shampoo, kerosene, milk, soft drinks, detergent, kitchen vegetable oil, wall, blackboard, cork, leather, wheat flour, corn flour, acrylic thread, salt, sugar, sawdust, earth or clay, brick, rubber, etc.

We can compare the results of this experiment with Experiments 6.12 and 6.13. The conclusion is that the concepts of *conductors* and *insulators* are relative. That is, substances like glass and wood which behave as conductors in the usual experiments of electrostatics, now behave as insulators when under a potential difference of 1.5 V. We will not go into details here, but in electrostatics it is common to work with potential differences ranging from 1,000 volts to 10,000 volts (that is, between 10^3 V and 10^4 V). The potential difference referred to here is between the charged body (like a rubbed piece of plastic, or a charged electroscope) and the ground; or between the ends of a body (when we want to test if this body behaves as a conductor or as an insulator). In these cases, most substances behave as conductors, as we saw previously. The potential difference generated between the poles of common chemical piles or batteries, on the other hand, is much lower, ranging from 1 V to 10 V. For these low potential differences several substances like wood, paper, rubber, and glass, behave as insulators. This shows that we must be very careful in classifying the substances as conductors or insulators. After all, the behaviour of all substances depends not only upon their intrinsic properties, but also upon the external potential difference to which they are submitted. This is a very important aspect which should be always kept in mind.

There is a gradation between the usual experiments of electrostatics and the experiments for which there is a potential difference of few volts. In other words, there is a gradation in the conducting and insulating properties of bodies when we go from a potential difference of 10,000 V to a potential difference of a few volts.

Due to this fact, it might be appropriate to change our terminology. Normally we say that a certain body A *is* a conductor, while another body B *is* an insulator. However, from what has just been seen, it would be more correct to say that in a certain set of conditions body A *behaves as* a conductor, while in another set of conditions it *behaves as* an insulator. The same would be valid for body B. But this would make all sentences very long and complicated. For this reason we will maintain the usual procedure of saying that bodies *are* conductors or insulators. But it should be clear to everyone that these are relative concepts, which depend not only on intrinsic properties of these bodies, but also on the external conditions to which they are subject.

6.7 Other Aspects which have an Influence upon the Conducting and Insulating Properties of a Substance

Conductors were defined in Subsection 6.3.1 as the substances which discharge an electrified electroscope when we touch the electroscope with this substance. Insulators, on the other hand, were defined as the substances which do not discharge an electrified electroscope when we touch the electroscope with it. In Section 6.6 we saw that these are relative definitions. After all, depending upon the electric potential difference acting upon the ends of a substance, it can behave as a conductor or as an insulator. In this Section we will mention briefly other three aspects which also have an influence upon these definitions.

6.7.1 The Time Necessary in order to Discharge an Electrified Electroscope

Experiment 6.21

We charge an electroscope and let it on a table in a dry day. We observe that the strip remain raised for several seconds or even for a few minutes. However, if we wait for a sufficiently long time as one hour, for instance, the electroscope will be totally discharged.

This means that the definitions of a conductor and of an insulator of Subsection 6.3.1 depend upon the observation time. For a time interval of a few seconds, dry air can be considered as a good insulator. For a time interval of one hour, on the other hand, dry air can be classified as a conductor, as it allows the discharge of an electroscope.

Definitions: In the present definitions we are referring to the experimental procedures described in Section 6.3. For the purpose of this book, we can define *good conductors* as the substances which discharge an electrified electroscope when they get in touch with it during a time interval smaller than 5 seconds. *Bad conductors* or *bad insulators* are the substances which discharge the electroscope during a time interval ranging from 5 seconds up to 30 seconds. These bodies are also called *poor conductors*, *poor insulators*, *imperfect conductors* or *imperfect insulators*. Finally, *good insulators* are the substances which require a time interval larger than 30 seconds in order to discharge an electrified electroscope.

6.7.2 The Length of a Substance which Comes into Contact with an Electrified Electroscope

Experiment 6.22

We cut several strips of paper, 2 cm wide and with lengths varying from 10 cm up to 1 m. We charge an electroscope and let it on a table on a dry day. We hold an end of the 10 cm strip with our hand and touch its free end at the edge of the thin cardboard of the electroscope. We observe a quick discharge of the electroscope. By the definition of Subsection 6.7.1, this means that this strip can be considered as a good conductor.

We charge once again the electroscope and now utilize a 30 cm $\times$ 2 cm paper strip. With our hand we hold one end of the paper strip and its free end touches the cardboard of the electroscope. Now we can easily notice the required time interval of a few seconds in order to discharge the electroscope. Depending upon the type of paper, this 30 cm long strip of paper may be considered as a bad conductor.

This experiment also shows clearly that, with the passage of time, increases the amount of charge lost by an electrified electroscope, see Subsection 6.7.1. The only difference is that in the present situation the electroscope is being discharged mainly through the paper strip and not through the surrounding air.

The electroscope is charged once more and the experiment repeated with a 1 m long and 2 cm wide paper strip. We observe that the electroscope remains charged for several seconds. By the definition of Subsection 6.7.1, this means that this 1 m long paper strip may be considered as a good insulator.

This experiment shows that the length of a substance has an influence upon its behaviour as a conductor or as an insulator. By increasing the length of a substance between our hand and the cardboard of the electroscope, we increase the amount of time required to discharge the electroscope.

6.7.3 The Cross-Sectional Area of a Substance which Comes into Contact with an Electrified Electroscope

Experiment 6.23

We charge an electroscope and let it on a table. We hold the end of a single human hair in our hand and touch the other end of the hair in the thin cardboard of an electrified electroscope. We observe that the electroscope remains charged for several seconds. This indicates that we can classify a single human hair as a good insulator.

We charge the electroscope once more and increase the number of hair threads in our hand, with their free ends touching simultaneously the cardboard of the electroscope. We observe that by increasing the number of hair threads, the discharge becomes faster. For instance, with dozens of hair threads the electroscope discharges in a few seconds. We can then classify this amount of hair as a good conductor.

Experiments like this one show that the cross-sectional area of a body has an influence upon its conducting or insulating properties. The larger this area between our hand and the cardboard of the electroscope, the smaller will be the time required to discharge the electroscope.

In this book we will not discuss in greater details the aspects presented in Sections 6.6 and 6.7.

6.8 Electrifying a Conductor by Friction

Experiment 6.24

We saw in Experiment 2.11 that we cannot charge a metal by friction while holding it with our hands. Now that we have discovered the distinction between conductors and insulators, together with the fact that the human body is a conductor, we can try a variation of this experiment. After all, it is possible that the metal had acquired a charge when we rubbed it, but this charge would have immediately been discharged to the ground through our body. With this new experiment we conclude that this was indeed the case.

This time we support the metal at the end of a PVC tube 30 cm long. The metal might be, for instance, an aluminum kettle. The kettle is supported upside down through its open mouth around the PVC tube. We charge an electric pendulum negatively and another one positively, as in Section 5.4. These two charged pendulums are kept separated from one another.

We hold the tube with our hands, without touching the metal. We wrap a plastic bag on the other hand and rub this plastic against a section of the kettle. Still holding the PVC tube with our hands and without touching the metal, we then bring the rubbed section of the aluminum kettle near the two oppositely charged pendulums, as always preventing these substances from coming into contact with the charged paper disks of the pendulums. We then bring the rubbed section of the plastic bag near the two oppositely charged pendulums, once more preventing them from coming into contact. By the attraction and repulsion exerted between these substances and the pendulums we will find that the plastic has become negatively charged, while the aluminum kettle has become positively charged.

Experiment 6.25

We repeat the procedures of Experiment 6.24. But we now consider an unrubbed portion of the plastic bag which is far away from the rubbed portion. When we move the unrubbed portion near the two charged pendulums, the pendulums will not indicate the presence of a net charge on this unrubbed section of the plastic. On the other hand, any portion of the insulated kettle will affect the pendulums when the kettle is brought near them. This happens no matter whether the portion of the kettle brought near the pendulums is close or far away from the rubbed portion of the kettle. That is, all portions of the kettle will behave as being positively charged.

Experiments like this show that we can electrify a conductor by friction, provided the conductor is insulated during this procedure.

The discovery that metals can also be charged by friction, provided they are well insulated, was not made until the 1770's.[6]

6.9 Conservation of Electric Charge

Experiment 6.26

We take a neutral PVC tube, together with a piece of equally neutral plastic bag, as in Experiments 2.1, 3.1 and 4.5. We charge an electric pendulum negatively and another one positively, as in Section 5.4. We then rub the plastic bag on the PVC tube. We now slowly bring the rubbed portion of the tube near each one of the charged pendulums, preventing contact between the tube and both pendulums. We can then conclude that it has become negatively charged, as it repels the negative pendulum and attracts the positive one. By slowly bringing the rubbed section of the plastic bag near each one of the charged pendulums,

[6] [Hem80] and [Hei99, p. 252, note 10].

as always preventing them from coming into contact, we can conclude that the plastic bag has become positively charged.

Let us now analyze the experiments of Section 5.4, together with Experiments 6.24 up to 6.26. They show that when we rub two initially neutral bodies together, one of them acquires a positive charge and the other a negative charge. This can only be easily detected when both bodies are insulators. When one of these bodies is an insulator and the other a conductor, normally the charge acquired by the latter will be discharged to the ground through our body. This may give the impression that the charge on the insulator arose out of nothing, as the conductor became discharged after friction. In order to observe the charge acquired by a conductor during friction with another body, it is essential to electrically insulate the conducting body. When it is insulated, the charge it has acquired will not be discharged to the ground, allowing its properties to be observed.

Experiment 6.25 also again shows that the charge acquired by an insulator due to friction does not distribute itself throughout the insulator. That is, it remains attached to the rubbed region. The charge acquired by a conductor due to friction, on the other hand, quickly distributes itself across the conducting surface. It does not remain attached to the rubbed region.

In the next experiments we utilize two electroscopes of the same size. We also suppose that they have strips of the same length and width, and equally sensitive (that is, with the same angular mobility in relation to the vertical). Normally we will also consider that they are equally charged. This can be indicated by the angle of the strips in relation to the vertical. After they have been charged, we must not touch them with our body to avoid discharging them. When we need to move them, it is best to hold them by the plastic supports.

Experiment 6.27

We negatively electrify two electroscopes with the same amount of charge, as in Experiment 6.2. We place them side by side in the same plane, separated by a small distance. After the two strips have been raised and the rubbed straw which charged them has been removed, we touch the two rectangles against one another. We observe that the two strips remain raised (Figure 6.29). Both of them remain raised after the two electroscopes have been separated. The same happens when we electrify the two electroscopes with equal positive charges.

Experiment 6.28

We charge one electroscope negatively, as in Experiment 6.2, and the other electroscope positively, as in Experiment 6.8. After the strips have been raised, we remove the two rubbed straws which charged the electroscopes. The two electroscopes are placed in the same plane side by side, at a small distance from one another. We then touch the two electroscopes. This time the two strips

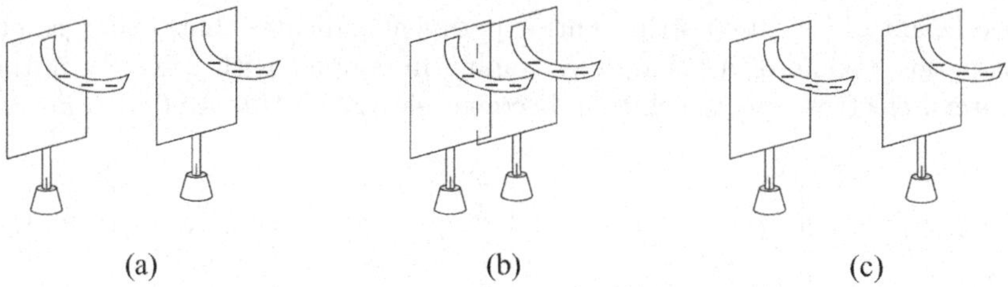

Figure 6.29: (a) Two electroscopes electrified with charges of the same sign. (b) The cardboards touch one another and nothing happens. (c) After separation, the strips remain raised.

immediately drop, returning to their natural vertical orientation (Figure 6.30 (b))! We can separate the two electroscopes and the strips remain vertical. This indicates that the two electroscopes, initially electrified with opposite charges, have been discharged by mutual contact. This experiment is the opposite of Experiments 6.24 and 6.26. In these last experiments we had two bodies which were initially neutral. After the experimental procedures they became electrified with charges of opposite sign. In the present experiment, on the other hand, we have two electroscopes electrified with charges of opposite sign. After the experimental procedure they became neutralized.

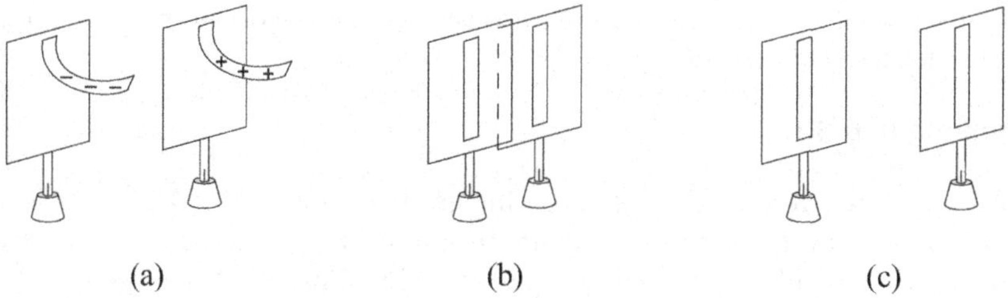

Figure 6.30: (a) A positive electroscope and a negative electroscope. (b) After contact, the strips drop. (c) After separation, the strips remain vertical, indicating that the electroscopes are now discharged.

Experiment 6.29

A single electroscope is negatively charged, as in Experiment 6.2. It should have a good amount of charge. This is indicated by a large angle of inclination of its strip in relation to the vertical (Figure 6.31 (a)). We place another electroscope, initially discharged, at the same vertical plane, at a small lateral distance from the first one. We then make the two electroscopes touch one another. We observe that the angle of the strip on the first one decreases, while the strip on the other electroscope rises (Figure 6.31 (b)). After separation, the strips remain unchanged. The amount of charge of an electroscope may be indicated by the height of its raised strip. This experiment indicates that a charged electroscope loses some of its charge by coming in contact with a second initially uncharged electroscope, which then becomes electrified (Figure 6.31 (c)).

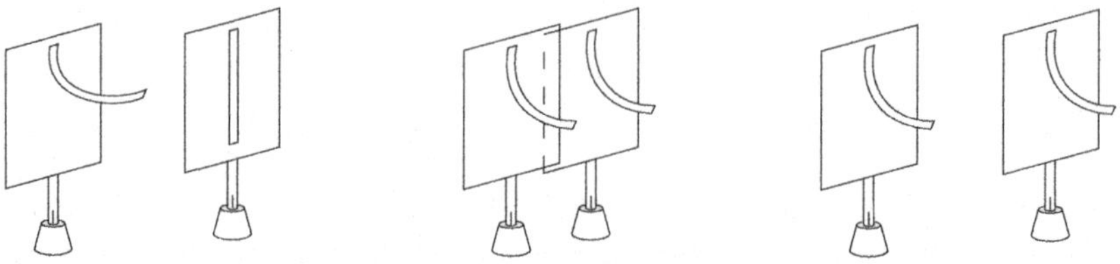

(a) (b) (c)

Figure 6.31: (a) A charged and an uncharged electroscope. (b) Contact between the cardboards. (c) Separation after contact. The strips remain raised, but with an inclination smaller than the inclination of (a).

When we separate these two electroscopes, they both remain weakly charged. By slowly moving a negative straw near both strips, we can conclude that both electroscopes are now negatively charged. The inclination of the first electroscope decreased during this experiment, while that of the second one increased. This suggests, by Experiment 6.9, that part of the original negative charge of the first electroscope was transferred to the second electroscope.

The same effect happens between a positively charged electroscope and an initially discharged electroscope.

Experiment 6.30

We can build a variation of Experiments 6.14 and 6.15. Electroscopes *I* and *II* remain at a distance of some 15 cm from one another. The strips remain on the outside. The electroscopes should be initially discharged, without any connection between them in this experiment. After this setup, we charge electroscope *I* by the procedure of Experiment 6.2. The strip on electroscope *I* will rise, while the strip on electroscope *II* does not move (Figure 6.32 (a)). After this procedure, the charged straw is removed. We take now a second, neutral straw. It should be placed above the cardboards of both electroscopes, and be supported by them. After it is placed there, nothing changes. The strip on electroscope *I* remains raised and the strip on electroscope *II* points downward (Figure 6.32 (b)).

Experiment 6.31

Experiment 6.30 is repeated. Electroscope *I* is charged while there is no connection between the electroscopes (Figure 6.33 (a)).

But now the two electroscopes will not be connected by a neutral straw. Instead of this, a plastic straw is cut so that its length is about 10 cm. A wood skewer or a copper wire, 20 cm long, is passed through it. With our hand, we

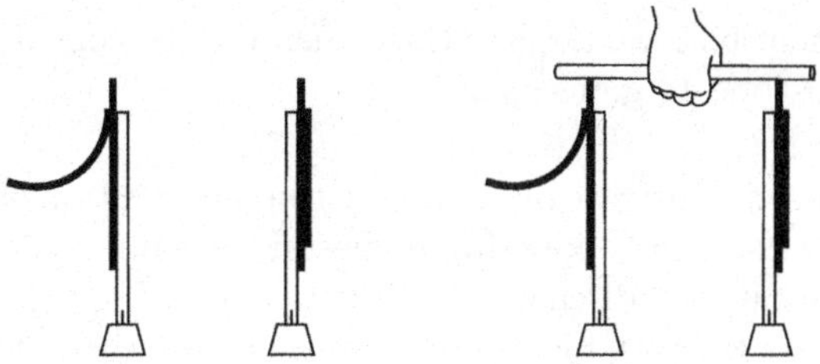

(a) (b)

Figure 6.32: (a) Electroscope I is charged. (b) A neutral straw is then placed in such a way as to connect both electroscopes. Nothing changes when this is done.

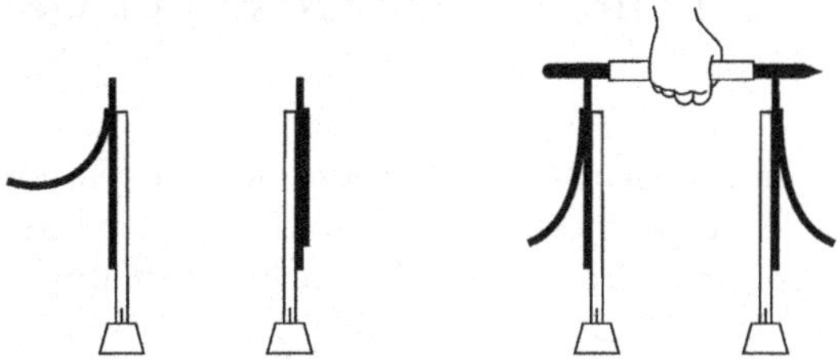

(a) (b)

Figure 6.33: (a) Only electroscope I is charged. (b) A wood skewer is surrounded by a plastic straw, with the wood supported by the two electroscopes. We observe that strip I falls a little, while strip II rises a little.

touch only the straw and place the ends of the skewer so it is supported by the cardboards. Now the strip on electroscope I drops a little, while the strip on electroscope II rises a little (Figure 6.33 (b)). At the end of the procedure both strips will be at the same height, but lower than the situation of Figure 6.33 (a).

Two precautions are essential for the success of this experiment. The first is that we cannot touch the cardboards, the wood skewer or metal wire with our hands. The second is that the skewer or wire must be enclosed in two or three plastic straws, as not all straws are good insulators. As a result, we may discharge electroscope I partially or totally through our body by placing the skewer on the cardboards. The electric charges on electroscope I will go to the skewer, pass through the straw, and then discharge through our body. By placing two or three straws one inside the other, we increase the insulation of the system. If difficulty arises trying to put straws inside one another, any alternate insulator we have found may be used to support the wood skewer (see Subection 6.3.2).

In Experiment 6.30 there was no transfer of electric charge between electroscopes I and II when they are connected through a plastic straw. On the other hand, in Experiment 6.31 we observe that when electroscope I looses a little of its charge, electroscope II gains some charge.

These experiments illustrate qualitatively the conservation of electric charges. When two equally charged conducting bodies of the same size and shape touch one another, they are not discharged. When they are oppositely charged, on the other hand, the two conductors discharge one another, remaining neutral after contact. When only one of them is initially charged, the contact with the second conductor initiates a transfer of charge from the electrified body to the neutral one. At the end of this process both conductors become electrified with charges of the same sign.

6.10 Gray and the Conservation of Electric Charges

Perhaps the first to notice something like these effects experimentally was Gray in 1735. He suspended a boy by silk strings, so that the boy could lay in a prone position. At one side of the boy there was a man standing on an insulating support made of gum-lac[7] and resin. At the other side of the boy there was a second man holding a "pendulous thread." This was a kind of electroscope invented by Gray. It was probably only a linen or cotton thread attached to a wooden stick. As this is a conducting thread, it is attracted by a nearby charged body (the boy in this example). The angle of inclination of the pendulous thread to the vertical would be an indicator of the amount of electrification of the nearby body. Gray had mentioned this kind of electroscope previously.[8] The second man in Gray's experiment was probably in direct contact with the ground, not being insulated. Gray rubbed a glass tube with his hands and brought it near the boy's feet. After this procedure, the boy attracted the thread of the electroscope which was held by the second man. The boy then moved his finger near the man standing on the resin. There was an electric discharge and the boy lost part of his attracting power (as indicated by a decrease in the angle of inclination of the electroscope). At the same time, the man standing on the resin has become electrified, attracting the thread of the electroscope.

We quote the relevant passages of this crucial experiment:[9]

> May the 6th [of 1735], we made the following experiment. The boy being suspended on the silk lines, and the [rubbed glass] tube being applied near his feet as usual, upon his holding the end of his finger near a gentleman's hand, that stood on a cake made of shell lack and black rosin; at the same time another gentleman stood at the other side of the boy with the pendulous thread; then the boy was bid to hold his finger near the first gentleman's hand, upon which

[7] See Appendix A.

[8] [Grad, p. 228], [Graf, p. 289], and [Grai, pp. 167-168]. See also Section 4.9.

[9] [Grai, p. 168].

> it was prick'd, and the snapping noise was heard; and at the same time, the thread which was by its attraction going towards the boy fell back, the boy having lost a great part of his attraction, upon a second moving his finger to the gentleman's hand, the attraction ceased; the thread being held near that gentleman, he was found to attract very strongly; but having since repeated this experiment, I find that though the attraction of the boy is much diminished, yet he does not quite lose it, till 2, 3, and sometimes 4 applications of his finger to the hand of him that stands on the electric body, but without touching him.

The idea of conservation of charge was implicitly present in the conceptions of several scientists who worked with electricity. With Gray we have the first experiment demonstrating this fact qualitatively. One of the first to explore fruitfully the concept of the conservation of charge was Benjamin Franklin (1706-1790) between 1745 and 1747.[10]

6.11 A Short History of the Electroscope and the Electrometer

The oldest procedure for determining whether a body was charged or neutral was to bring it near light substances, as in the experiment on the amber effect. Later, more sensitive instruments were devised to detect this property of bodies: Fracastoro's perpendiculo, Gilbert's versorium, and Gray's pendulous thread. In all these cases it was necessary to decrease the distance between the rubbed body and any of these instruments in order to observe how the instrument reacted to the presence of the body, be it a piece of amber or a plastic straw.

Usually there is no visible difference between a neutral body and a charged body. A piece of amber, for instance, does not change its color or its shape by being charged. In general we can only detect whether or not it is charged by the effects it causes upon nearby substances or a sensitive instrument placed near it. When an electric pendulum is far away from other bodies, it will hang vertically, whether it is charged or not. Only when we bring it near other substances can we detect, by the angle of the string of the pendulum to the vertical, if the pendulum is charged or neutral.

In this book we call an instrument which, when connected to a body, automatically indicates whether or not this body is charged, an electroscope. It should indicate this electricity automatically, in such a way that we do not need to touch the electroscope or to carry out any other procedure. Even when this instrument is not connected to any other body, it is possible to know if the electroscope itself is charged or not. In this regard, the electroscope is a different instrument from the versorium and also from the electric pendulum.

An instrument must have two main properties in order to be characterized as an electroscope. (I) The first is that the electroscope must be electrically

[10][Hei99, pp. 327-333].

insulated from the ground. As an alternative, the body to which the electroscope is attached must be insulated from the ground. This electrical insulation is essential. Without this property, the electroscope cannot preserve its acquired charge. As regards the electroscopes utilized in this book, this insulation is accomplished by supporting the thin cardboard by the neutral plastic straw. (II) The second property is that the electroscope must have a visible part which changes its state depending whether the electroscope is neutral or charged. In the electroscopes in this book, this property is the angle between the mobile paper strip and the stationary thin cardboard. In the classical electroscopes, this state is indicated by the angle between the two mobile straws.

Perhaps the first electroscope with these two properties was built by Du Fay in 1737.[11] Gray, before Du Fay, had already employed a cotton or linen thread attached to a stick in order to indicate whether a nearby body was charged. But in order to indicate the charge of the nearby body, it was necessary to move the stick near the body and observe whether the thread inclined toward the body. Du Fay, on the other hand, began to hang threads folded in the middle on the body. When this body was charged, the two halves of each thread would then move away from one another, making an upside down letter V. The angle of separation between the two halves would increase with the amount of charge on the body. In the first experiment in which he described this new procedure,[12] he suspended a iron bar horizontally by silk cords. These cords insulated the bar from the ground. On the bar he suspended several threads of the same length but made of different substances: linen, cotton, silk, and wool. Each thread was folded in the middle, with their central points supported by the bar. After electrifying the bar, he noticed that the two halves of each thread moved away from one another. Moreover, the angle of separation was greater for the linen thread, followed by cotton, silk, and wool, which had the least angular separation. He believed that this fact indicated that linen had a greater capacity for accumulating electrical matter than the other substances.

He then presented the following extremely interesting description:[13]

> A length of [cotton or linen] thread placed above an iron bar sus-

[11] [DF37b, pp. 94-98].

[12] [DF37b, pp. 95-96].

[13] [DF37b, p. 98]. Une aiguillée de fil posée sur une barre de fer suspenduë par des cordons de soye, présente l'idée de la plus simple de toutes les expériences, cependant elle peut fournir de sujet à des méditations profondes, & elle sert à confirmer la plûpart des principes que j'ai établis dans mes Mémoires précédents, tant sur la communication de l'électricité & ses effets de répulssion & d'attraction, que sur la réalité des deux genres d'électricité, sçavoir la vitrée & la résineuse. Elle sert aussi à connoitre si la force de l'électricité est plus ou moins grande, ce qui est très-commode dans la pratique de toutes ces expériences; il ne s'agit pour cela que de poser sur la barre le bout de fil, comme nous l'avons dit, on verra pour lors les deux bouts qui pendent librement d'un côté & de l'autre de la barre s'écarter l'un de l'autre avec plus ou moins de force, & former un angle plus ou moins grand, suivant que la barre aura reçû du tube plus ou moins de vertu électrique, & cela fera connoître d'une maniére assés exacte, le degré de force de l'électricité, de sorte que l'on pourra choisir le temps & les circonstances les plus favorables pour les expériences qui demandent la plus forte électricité, telles que sont celles qui concernent la lumiére, ou la communication le long d'une corde ou d'un autre corps continu.

pended by silk cords, presents the simplest idea of an experiment, although it offers the subject of profound meditations, and it serves to confirm the major part of the principles which I established in the previous works, not only about the communication of electricity and its effects of attraction and repulsion, but also about the reality of the two kinds of electricity, namely, the vitreous and resinous. It can also be utilized in order to know if the electric force is greater or smaller, which is very useful in the practice of all experiments. As we said before, to do this, it is only necessary to place on a bar the piece of linen thread. It will be seen the two ends which hang freely in both sides of the bar, move away from one another with a bigger or smaller force, making a greater or smaller angle, depending whether the bar received from the [rubbed glass] tube a greater or smaller electric virtue. [An illustration of an experiment of this kind appears in Figure 6.34.] And this will make known in a very precise way the degree of the force of electricity, in such a way that we will be able to choose the moment and the most favorable circumstances to perform the experiments which require the strongest electricity, as are the experiments related with light, or the communication along a string or along another continuous body.

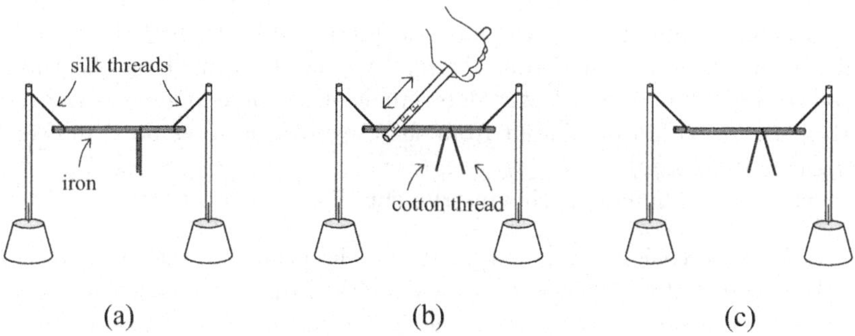

Figure 6.34: Du Fay's electroscope. (a) Iron rod supported by silk cords. A cotton thread is suspended by the bar. (b) A rubbed glass tube is scraped against the bar. (c) When we remove the glass, the two halves of the cotton thread move away from one another.

Du Fay's experiment is analogous to our Experiment 6.9, represented by Figure 6.12. By increasing the amount of charge upon the electroscope, we increase the angle of separation of the leaves.

Du Fay also utilized his electroscope to discover which bodies are the best insulators. To do this, he suspended an iron bar by strings made of different substances, or supported the iron bar on solid bodies made of different materials. After this preparation, he charged the iron bar with a glass tube that had been previously rubbed. He then noted, by observing the angle of his electroscope connected to the bar, which substance allowed the bar to remain charged for a longer time. This substance (the strings by which the bar was suspended, or the rigid support below the bar holding it above the ground) would then be the best insulator.[14]

Jean Antoine Nollet (1700-1770) (Figure 6.35), was Du Fay's assistant for a few years, from 1731 or 1732 to about 1735.[15]

Figure 6.35: Jean Antoine Nollet (1700-1770).

In 1747 Nollet presented an improvement on Du Fay's electroscope. The open threads, like the inverted letter V, were connected directly to the charged body. A lamp illuminated the threads and projected their shadows on a graduated screen where he could directly read the angle formed by the threads. This allowed great precision in the determination of these angles, as the screen and the observer could be distant from the electroscope so as not to affect the reading (Figure 6.36).[16]

Nollet created a name for this instrument: the *electrometer*:[17]

> In several occasions I utilized a very simple means, in order to know the progress of the electricity, which deserved the name *electrometer*, if it were generally employed, and if it could serve to measure by well known quantities, which we could not doubt, the increases or decreases which it indicates.

This is an appropriate name, as this instrument allows a quantitative measure of an angle. And the value of this angle is related to the electricity of the body to which it is connected. The electrometer is an electroscope with which we can make a precise quantitative measurement of a property—such as the angle in this case—associated with electricity.

---

[14][DF37b, p. 99].

[15][Hei81e] and [Hei99, pp. 279-289].

[16][Nol47, p. 129] and [Hei99, p. 353].

[17][Nol47, p. 129]: Dans bien des occasions je me suis servi, pour connoître les progrès de l'électricité, d'un moyen assez simple & qui mériteroit le titre d'*électromètre*, s'il étoit généralement applicable, & s'il pouvoit servir à mesurer par des quantités bien connues, & dont on ne pût douter, les augmentations ou diminutions qu'il indique.

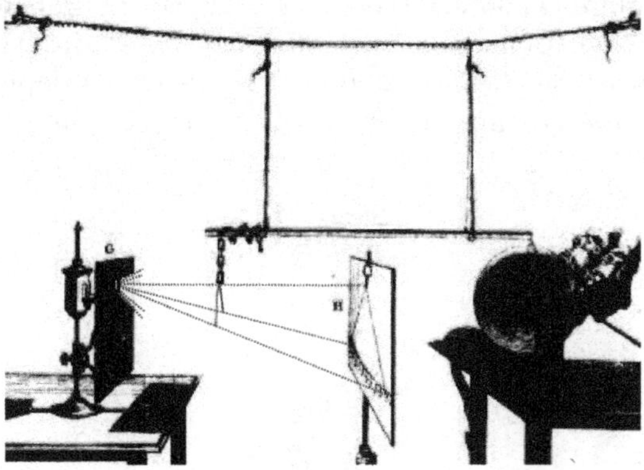

Figure 6.36: Nollet's electrometer.

Another example of one of his electrometers is given in Figure 6.37.[18]

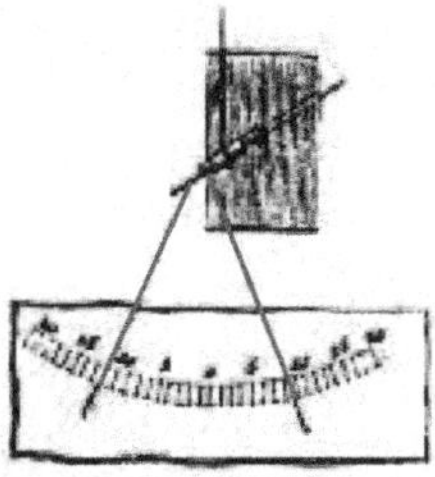

Figure 6.37: Another electrometer made by Nollet.

To prevent electricity losses from the ends of the threads of any electroscope, later on the threads were terminated in small, light balls made from elder pith or cork. One of the scientists responsible for this development was John Canton in 1752-4.[19] Figure 6.38 presents Canton's electroscope.

Figure 6.38: Canton's electroscope.

Later on the linen threads were replaced by rigid straws and metal leaves.

[18][Nol67, Planche 4, Figure 15].
[19][Can53], [Can54], and [Wal36].

They were more durable and allowed a more precise determination of the aperture angle. Abraham Bennet (1750-1799) and Alessandro Volta (1745-1827) were important scientists who contributed to these developments. Bennet's electrometer with its two slips of gold leaf is represented in Figure 6.39.[20]

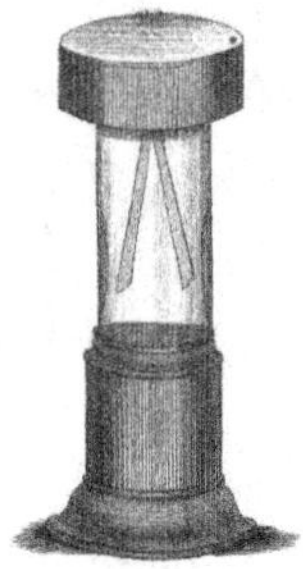

Figure 6.39: Bennet's electrometer.

Georg Wilhelm Richmann (1711-1753) created an electrometer between 1744 and 1753 in which one of the straws is fixed relative to the ground and only a single straw or strip moved away from the vertical when the electroscope was charged (Figure 6.40).[21] This is analogous to the electroscope of Figure 6.1 which is used in most experiments in this book.

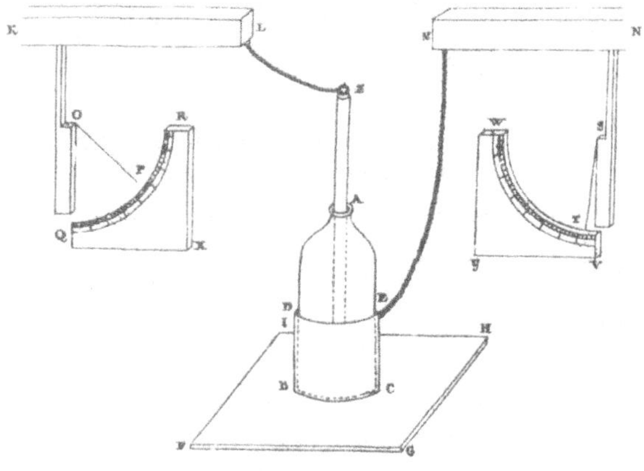

Figure 6.40: Richmann's electrometers connected to a Leyden jar. Each one of the electrometers has a single mobile strip.

This kind of electrometer was developed later by William Henley (date of birth unknown, died in 1779) in 1772 (Figure 6.41).[22] It has a single light mobile

[20][Ben86] and [Hei99, p. 450].

[21][Hei99, p. 392].

[22][Pri72].

rod A, with a cork ball at the end, made to turn on the center B. When the electrometer is charged, there is a repulsion between the mobile rod A and the fixed stem C. The opening angle can be read at the protractor connected to the instrument.

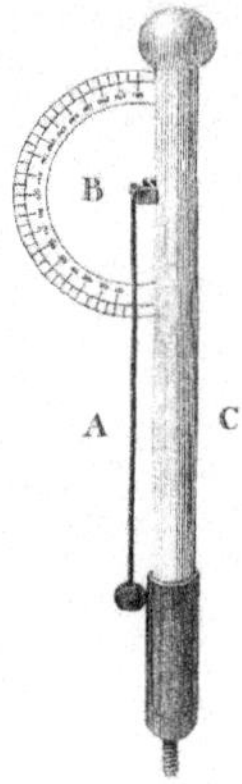

Figure 6.41: Henley's electrometer.

For more details about the history of the electroscope and electrometer, see the works of Walker, Heilbron, and Medeiros.[23]

[23][Wal36], [Hei99, pp. xvi, xx, 82, 259, 327, 331, 353, 367, 373-376, 390-392, 418, 421-422, 447-456, 462, and 491-494], and [Med02].

Chapter 7

Differences between Conductors and Insulators

7.1 Mobility of Charges on Conductors and Insulators

We saw that the main property of an insulator like plastic is that it does not allow a flux of electric charges through it. Therefore, it does not discharge an electrified electroscope. A conductor, on the other hand, allows the passage of charges through it. Examples of conductors are the human body, the ground, a metal, a sheet of paper or the thin cardboard of an electroscope. Therefore, when an electrified electroscope is connected to the ground through a conductor, it is discharged. We will now look at other properties which distinguish conductors and insulators.

Experiment 7.1

We cut a rectangular strip of thin cardboard, 30 cm long and 2 cm wide. The plane of the strip will be vertical, with its longer side parallel to the horizon. We attach the upper end of a vertical plastic straw to the center of the strip, forming a letter T. The lower end of the straw should be fixed to an appropriate support, like modeling clay or the support of the electric pendulum, as in Section 4.4. We then hang four very thin strips of tissue paper, 10 cm long, spread along the thin cardboard strip. They should be folded in the middle, in such a way that both halves hang side by side vertically. After this procedure, we rub another plastic straw. It should be scraped one or more times over the horizontal thin cardboard, charging it. We remove the rubbed straw. All strips of tissue paper open to the outside, with each half moving away from the other half.

The same experiment can be repeated with a larger thin cardboard strip, for instance, 60 cm long. To do this, we can fix two vertical plastic straws equally spaced, one at a distance of 20 cm from one end of the cardboard, and the other at a distance of 20 cm from the other end. We spread several very thin strips of tissue paper over the upper horizontal edge of the wide thin cardboard strip, like several electroscopes. When we charge the thin cardboard by scraping it with a rubbed straw, all thin strips of tissue paper open their legs. Instead of a rectangular thin cardboard strip we can also use a hard copper wire.

We now build a T of the same dimensions, but made only of plastic. The horizontal upper portion of the T can be a plastic ruler (with its plane vertical), or a series of plastic straws attached to one another by their adjacent ends. We hang the thin strips of tissue paper over the upper section of the T, spread along its length. We rub another plastic straw and scrape it over the upper edge of the plastic T. In this case the thin strips of tissue paper do not open their legs, except those close to the scraped region.

Experiment 7.2

We cut a disk of thin cardboard with a 20 cm diameter. The plane of the disk will remain horizontal, supported by plastic straws placed vertically at appropriate locations beneath it. With a pair of scissors or single hole punch-pliers we make many pairs of holes along a diameter of the disk, with the two holes of each pair very close to one another. We then pass a single thin tissue paper strip through each pair of holes in such a way that it hangs through the middle with each half passing through a hole and facing the other half. Another possibility is to attach the strips like a letter L glued on the lower side of the disk. Two strips side by side will build a letter T, with the vertical portion of the letter composed of two strips, hanging vertically side by side from the disk. We rub another plastic straw and scrape it along an edge of the disk. The strips of tissue paper open their legs. This also happens for those strips which are located far away from the region that was scraped.

The same effect does not happen for a plastic disk. In this case the strips of tissue paper which are far from the scraped region do not open their legs after scratching the plastic disk with a rubbed straw. Only those close to the scratched area will separate from one another.

These experiments show that when we charge a conductor, the charges tend to spread across its surface. In an insulator, on the other hand, they do not move freely along its surface, staying attached to the place where they were generated or transferred to the insulator. The same effect had been seen in Experiment 6.25.

7.2 Charge Collectors

Our next electric instrument is a charge collector. It is used to obtain a small amount of charge from any region of an electrified body. By moving this collected charge near previously positively and negatively charged pendulums, or previously charged electroscopes, it is possible to determine the sign of the collected charge. The magnitude of the attractions and repulsions generated on the pendulums and electroscopes also indicates whether the collected charge is large or small. Although an electric pendulum or an electroscope can also be employed as a charge collector, we will use this term to designate instruments built specifically for this purpose. Another possible use of a charge collector is to serve as a charge transporter between two spatially separated conductors.

The simplest collector is a ball of aluminum foil suspended at the end of a silk or nylon thread, with the other end tied to a plastic straw (Figure 7.1). The ball can also be made of paper or another appropriate conductor. As silk is an insulator, it prevents any collected electricity from discharging. By increasing the diameter of the sphere, we increase the amount of charge it will collect in any experiment.

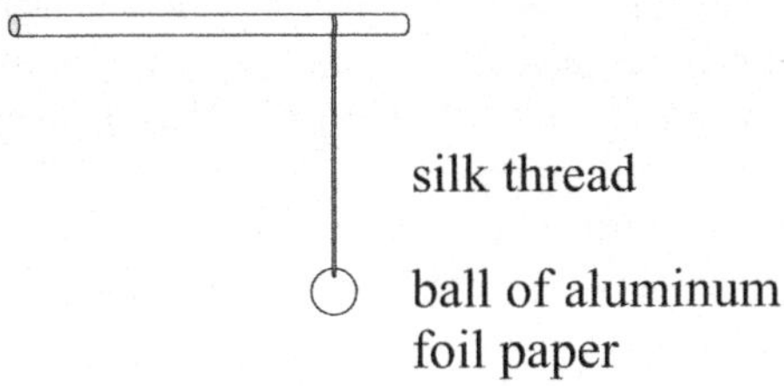

Figure 7.1: A charge collector.

A second model analogous to the previous one is composed of an aluminum foil ball atached to the end of a plastic straw (Figure 7.2). The main difference from the previous model is the rigidity of the straw. For this reason, it can be used to collect charges on the top or side portions of charged conductors, by holding the straw from below or sideways. That is, we have better control over where we can carry the aluminum ball.

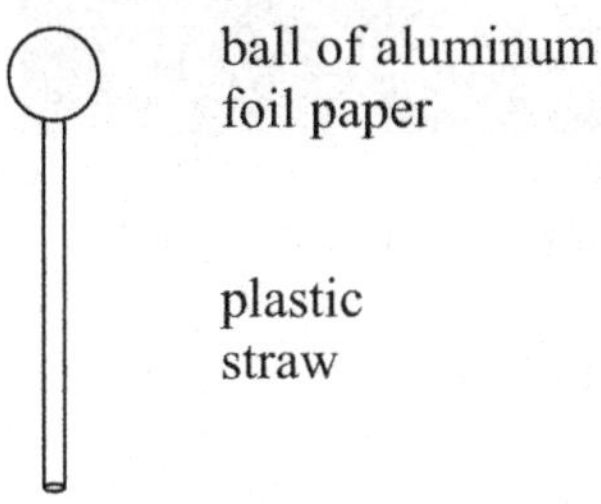

Figure 7.2: Another charge collector.

Another model is composed of a strip of aluminum foil attached to the end of a plastic straw. This strip may be, for instance, 5 cm long and 2 mm wide. We glue it around the tip of the straw (Figure 7.3).

Perhaps the oldest charge collector built exclusively for this purpose was made by F. U. T. Aepinus (1724-1802). There is no known portrait of Aepinus.[1]

[1][Aep79, p. 62].

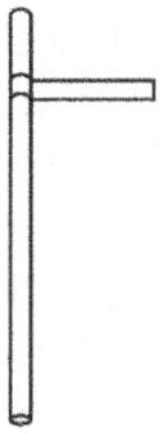

Figure 7.3: A third model of charge collector.

His charge collector was simply a small metal piece about 3.8 cm long fitted in the middle with a little hook to which he attached a well-dried silk thread. He described this instrument and several interesting experiments made with it in a book published in Latin in 1759, *Essay on the Theory of Electricity and Magnetism.*[2]

Another model was invented by Charles Augustin Coulomb (1736-1806) in 1787, and was called a proof plane (Figure 7.4).[3] Coulomb knew Aepinus's work and quotes it in his papers.

Figure 7.4: Charles-Augustin Coulomb (1736-1806).

The proof plane is a conducting disk attached by its center to an insulating handle. Coulomb used it to determine the distribution of charge on the surfaces of two or three conductors charged by contact. The amount of charge collected by the proof plane is proportional to the local surface density of charge. The model we use here is a thin cardboard disk 3 cm in diameter. We can place aluminum foil on one of its faces, but this is not essential. We cut a piece of plastic straw 5 cm long. It will be attached at right angles to the center of the disk, as if it were the axis of symmetry. One of the ends of the straw can be attached to the center of the disk with modeling clay (Figure 7.5). When we manipulate the proof plane, we must touch only the straw, but not the clay or

[2][Aep79, pp. 312-314].
[3][Hei99, p. 495].

the disk.

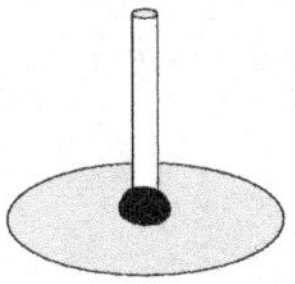

Figure 7.5: Coulomb's proof plane. This is also a collector but will be referred to as a proof plane for clarity when describing the following experiments.

These charge collectors will be used in some of the experiments described here.

7.3 The Electric Polarization of Conductors

Experiment 7.3

We saw in Experiment 6.12 that thin cardboard is a conductor, while plastic is an insulator. We use this fact to build an insulated conductor: A horizontal thin cardboard disk 15 cm in diameter is supported by four vertical plastic straws attached to appropriate supports, such as the supports of the electric pendulums. We place three Coulomb proof planes on the top of the disk, aligned along a diameter of the disk, one at the center and the other two close to the opposite edges. We call them 1, 2, and 3, with the 2nd proof plane at the center of the disk. Initially we touch the disk with our finger in order to discharge it. We now begin the experiment to study the distribution of charges on this disk in the presence of other nearby charged bodies. To do this, we first charge two electroscopes, one positively and the other negatively. We place them far away from one another and from this horizontal disk. We also will utilize a third discharged electroscope.

We charge a straw negatively along its entire length when we rub it in our hair. It should then be stood vertically on an appropriate support. The middle of this straw should be at the same height as the horizontal plane of the disk. The rubbed straw is brought near the disk, without touching it, close to proof plane 1 (Figure 7.6). They will then be in the following sequence: negative straw, proof planes 1, 2, and 3, respectively. The negative straw should be about 2 cm from the closest edge of the disk. We remove proof plane 2 and bring it close to the discharged electroscope. Nothing happens, indicating that it is electrically neutral. We can put it back in its original place. We now remove proof plane 1 and slowly bring it near the discharged electroscope, without making contact. The strip of this electroscope is attracted by this proof plane, indicating that the proof plane is charged. We then slowly bring it near the negative and positive electroscopes, as always preventing them from coming into contact. The charged proof plane attracts the strip on the negative electroscope and repels the strip on the negative electroscope. From these facts we conclude that proof plane

1 became positively charged due to the presence of the nearby negative straw. We then place it back in its original location above the horizontal disk. We now remove proof plane 3 and repeat these procedures, concluding that it has become negatively charged.

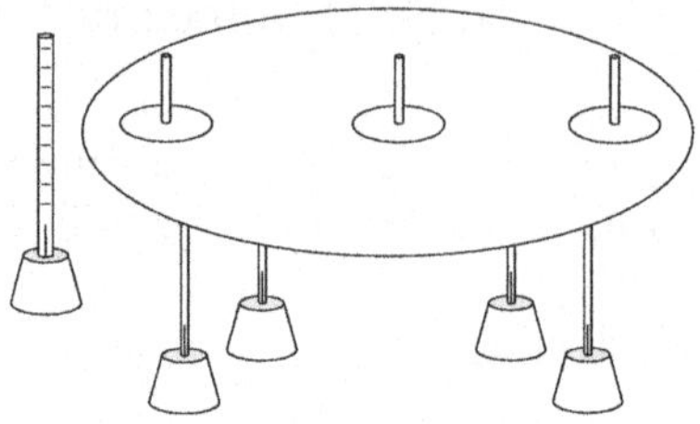

Figure 7.6: Experiment to show the distribution of charges on a conductor in the presence of nearby electrified bodies.

Experiment 7.4

We repeat Experiment 7.3 but now place the negatively charged straw about 5 cm from the closest edge. Once more this negative straw is aligned with Coulomb's proof planes in the following order: negative straw, proof planes 1, 2, and 3, respectively. When we perform the previous procedure, we find no charge on Coulomb's proof plane 2. Proof plane 1 is again found to be positively charged, but with a smaller amount of charge than proof plane 1 of Experiment 7.3. This is indicated by the attractive and repulsive forces it exerts upon the negative and positive electroscopes, respectively. These forces are of lower intensity in the present experiment than the analogous forces exerted by proof plane 1 in Experiment 7.3. After returning this first proof plane to its original place above the disk, we remove proof plane 3 and test its charge. Once more it is found to be negatively charged, but also with a smaller amount of charge than proof plane 3 of Experiment 7.3. This is indicated by the smaller repulsive and attractive forces it exerts upon the negative and positive electroscopes, respectively.

We can repeat Experiment 7.3, each time placing the negative straw farther from the disk. The farther it is, the smaller the amount of opposite charges collected by proof planes 1 and 3. When the negative straw is 20 cm away from the closest edge of disk, or even farther away, no detectable charge is collected by the proof planes in these experiments.

Experiment 7.5

We repeat Experiment 7.3. We will observe the forces exerted by the charged proof planes on the neutral, positive and negative electroscopes when the negative straw is about 2 cm from the edge of the disk.

We then put 2 or 3 negatively charged straws together, next to one another, with each straw on its own support. Their ends can also be tied together and all of them placed side by side on one support. This set of 2 or 3 straws should be put again about 2 cm from the edge of the disk. The straws should have approximately the same amount of charge, as they were rubbed in hair an equal amount of time. We repeat Experiment 7.3 and observe that proof plane 1 has again become positively charged, as before. But now it exerts a much greater force of attraction upon the strip of the neutral electroscope than the attractive force exerted by proof plane 1 of Experiment 7.3. It also exerts a greater attractive force upon the strip of the negative electroscope, and a greater repulsive force upon the strip of the positive electroscope. For this reason, we conclude that this proof plane has a larger amount of charge than the amount of charge collected by proof plane 1 in Experiment 7.3. The intensities of the forces exerted by proof plane 3 upon the strips of the electroscopes in this experiment are also greater than the analogous forces exerted by proof plane 3 in Experiment 7.3. Accordingly, we conclude that it acquired a larger amount of negative charge than proof plane 3 in Experiment 7.3.

Experiment 7.6

We cut a thin rectangular cardboard with sides of 10 and 7 cm. It will be set in a vertical plane with the longer side horizontal and the shorter side vertical. We attach a vertical straw to the center of the rectangle with adhesive tape. The lower end of the straw is attached to an appropriate support. We then touch the rectangle to discharge it. As in Experiment 7.3, we prepare a positively charged electroscope and a negatively charged electroscope beforehand. In this experiment we use the aluminum foil strip attached to a plastic straw as a collector of charge, as in Section 7.2 (Figure 7.3). We charge another straw negatively and attach it vertically to an appropriate support. The two charged electroscopes, the rectangle, and the charged straw are initially well separated from one another.

We now move the charged straw near to one of the vertical edges of the rectangle, without contact. While they are very close to one another, 1 or 2 cm away, we scrape the aluminum foil strip of the collector up and down over the other vertical edge of the rectangle (Figure 7.7). We then bring this strip near the two charged electroscopes, without allowing them to contact. By the attractions and repulsions observed in these electroscopes, we conclude that the strip of the collector has become negatively charged.

We discharge this strip by touching it with our finger. We now scrape it up and down over the vertical edge of the rectangle which is closest to the charged straw, taking care to not let it touch the straw. When we bring the strip slowly near the two charged electroscopes, we conclude that the strip has now become positively charged.

Experiment 7.7

We repeat Experiment 7.6, but now place the negative straw about 5 cm away from the closest edge of the rectangle. When we repeat this procedure, we again find that the side of the rectangle which is farthest from the negative

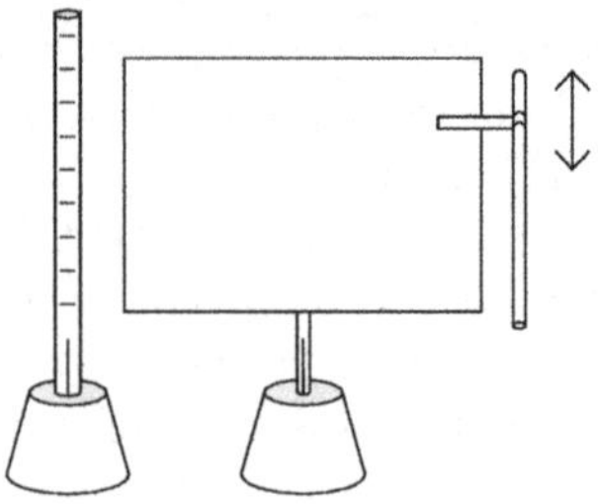

Figure 7.7: Studying the distribution of charges on a conductor.

straw has become negatively charged while the side of the rectangle which is closest to the negative straw has become positively charged. But the amounts of these collected charges are smaller than the analogous charges collected in Experiment 7.6. These amounts of charge can be estimated by the forces the charged strip exerts upon the positive and negative electroscopes.

When the negative straw is 20 cm away from the closest edge of the rectangle, or even farther from it, no appreciable amount of charge is collected by the metal strip.

Experiment 7.8

We repeat Experiment 7.6, but now place 2 or 3 negatively charged straws together as in Experiment 7.5. They should have approximately the same electrification, as they were rubbed equally in hair. They should be about 1 or 2 cm from one side of the rectangle. After repeating the same procedure as before, we find that the aluminum foil strip of the collector has acquired larger amount of positive and negative charges than the charges collected in Experiment 7.6.

These experiments show that the presence of the rubbed straw causes a separation of charges on the nearby conductor. The portion of the conductor which is closer to the rubbed straw acquires a charge of opposite sign to that on the straw, while the opposite portion of the conductor acquires a charge of the same sign as the charge on the straw.

Definitions: This phenomenon and this process are called electric or electrostatic *polarization*, *induction*, *influence*, *polarization by induction*, *polarization by influence*, *electrification by influence*, *electrification by induction*, or *electrification by communication*. In this work we utilize preferentially the first expression, *electric polarization*.

Experiments 7.4 and 7.7 show us even more. We can increase the amount of induced charges on both sides of the conductor by decreasing the distance between the conductor and the rubbed straw. This is represented in Figure 7.8.

Let us consider Experiment 4.5 again. The inclination of the pendulum relative to the vertical increases when the distance between the rubbed straw and the pendulum decreases. This indicates a greater force between them.

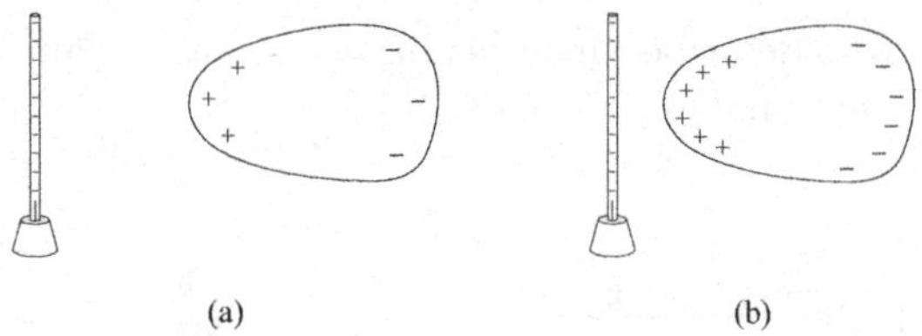

Figure 7.8: (a) The polarization of a conductor due to a nearby electrified insulator. (b) When we decrease the distance between these bodies, the amount of positive and negative charges induced on the conductor is increased.

From what has just been seen, we can increase the polarization of the disk by shortening its distance from the electrified straw (Figure 7.9).

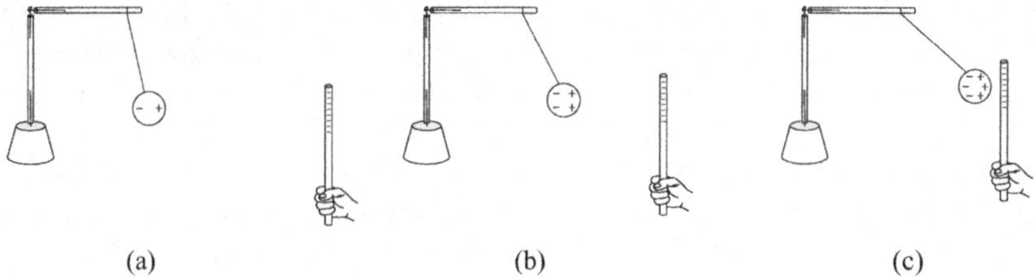

Figure 7.9: By decreasing the distance between a rubbed straw and a pendulum, we increase the polarization of the induced charges on the conducting disk.

Experiments 7.5 and 7.8 also show something relevant. We can also increase the amount of induced charges on both sides of the conductor by increasing the amount of charge on the nearby electrified insulator (Figure 7.10).

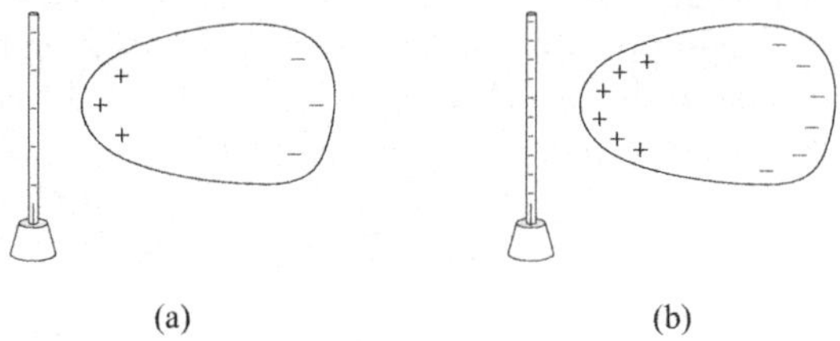

Figure 7.10: (a) The polarization of a conductor due to a nearby electrified insulator. (b) When we increase the electrification of the insulator, the amount of positive and negative charges induced on the conductor is increased.

7.3.1 Aepinus and Electric Polarization

One of the more important scientists who dealt with this topic was Aepinus in the period of 1755-1759.[4] An experiment analogous to Experiment 7.3 was first

[4] [Aep79], [Hei81a], and [Hei99, pp. 384-402].

done by Aepinus, who described it in his book of 1759.[5] Figure 7.11 presents a representation of one of his experiments.

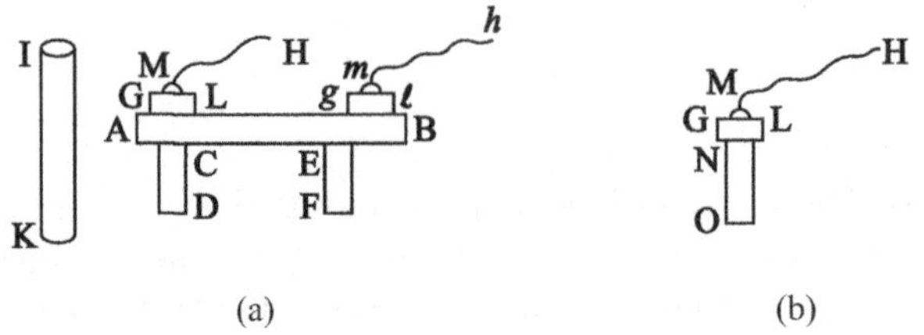

Figure 7.11: Aepinus's experiment to prove the polarization of conductors near electrified bodies.

Instead of a cardboard disk supported by plastic straws, he had a metal rod AB about 30 cm long on insulating glass supports CD and EF (Figure 7.11 (a)). His charge collectors have been described in Section 7.2. They were metal pieces GL and $g\ell$ about 3.8 cm long fitted in the middle with little hooks M and m to which had been attached well dried silk threads HM and hm. For charged bodies, instead of our negative rubbed straw, he used the electrificatory cylinder IK. It might be a glass cylinder electrified positively by rubbing, or a sulphur cylinder electrified negatively by rubbing.

He tested the charges induced in the ends of the metal rod due to the presence of each cylinder. He thus proved the polarization of the conductor. To do this, he took the cylinder electrified by rubbing, brought it close to the end A of the rod, to a distance of about 2.5 cm, and held it there motionless. He then lifted the metal piece GL by means of the silk thread HM and placed it on the glass support NO (Figure 7.11 (b)). When he brought positive and negative bodies close to his charge collector placed at NO, he concluded that it had acquired a charge having opposite sign to the electrificatory cylinder. When he performed the same test with the other charge collector $g\ell$, he concluded that it had acquired a net charge having the same sign to the electrificatory cylinder. That is, the ends A and B of the conducting rod acquired charges of opposite signs, with the charge of B being of the same sign to the electrificatory cylinder IK.

7.4 Attractions and Repulsions Exerted by a Polarized Body

Experiment 7.9

In this experiment we use the thin cardboard rectangle of Experiment 7.6, an electric pendulum and a plastic straw. We discharge the rectangle and the pendulum by touching them with our finger. After this procedure, we place

[5][Aep79, pp. 312-314].

them side by side along the same plane, with the pendulum close to right edge B of the plate. The section of the disk closest to the cardboard should be 2 to 5 cm distant from it (Figure 7.12 (a)).

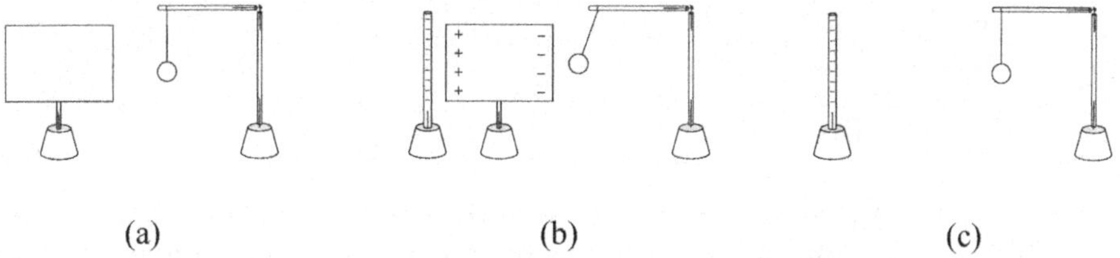

Figure 7.12: (a) A neutral pendulum hangs vertically close to a neutral conducting plate (left edge A and right edge B). (b) When a rubbed straw is brought near one edge of the plate, the pendulum is attracted by the other edge. (c) When we remove the plate, the pendulum returns to the vertical.

We electrify a plastic straw negatively by rubbing its entire length in our hair, attaching it vertically on a support. We then place it in the plane of the cardboard, opposite the pendulum, and far away from them. We then bring the straw slowly near left edge A of the plate. When it is close enough, the pendulum inclines toward the rectangle (Figure 7.12 (b)). We should not move the straw very close to the rectangle, in order to prevent contact between the disk and the rectangle. When we remove the straw, we observe that the pendulum returns to its vertical orientation.

We suppose the rubbed straw is again close to the rectangle, in such a way that the pendulum is inclined toward the cardboard, as in Figure 7.12 (b). We now remove the cardboard, without touching the straw or the pendulum. The rectangle should be removed in a direction perpendicular to its plane. After the rectangle has been removed, we observe that the pendulum returns to its vertical orientation (Figure 7.12 (c)).

This experiment shows that the pendulum is being attracted by the polarized plate and not by the rubbed straw, which is too far away from the pendulum. The electrified straw is responsible for the polarization of the conducting plate, but is too far away to noticeably affect the pendulum.

Experiment 7.10

In this experiment we use the thin cardboard rectangle of Experiment 7.6, an electric pendulum and a plastic straw. We charge the straw negatively by rubbing its entire length in hair, and then set it vertically on the support. When we move a neutral pendulum near the rubbed straw, we observe an attraction as they are brought very close to one another. This is indicated by the disk of the pendulum inclining toward the rubbed straw. On the other hand, when the distance between the rubbed straw and the paper disk of the pendulum is greater than or equal to about 15 cm, the pendulum remains vertical. Even though the rubbed straw may be attracting it, this force is so small that it is barely perceptible (Figure 7.13 (a)).

(a) (b)

Figure 7.13: (a) A negative straw at a distance of 15 cm from a neutral pendulum. (b) When we bring a conducting plate between them, we observe that the pendulum is attracted by the plate.

We now suppose that the rubbed vertical straw and the vertical thread of the neutral electric pendulum are separated by 15 cm, with the rectangular cardboard far away from them, as in Figure 7.13 (a). We then place the plane of the cardboard, with dimensions 10 by 7 cm, parallel to the plane connecting the straw and the pendulum's thread, with these two planes separated from one another. After this, we move the plate in a direction perpendicular to its plane, in such a way that it remains between the rubbed straw and the pendulum, all of them in the same plane (Figure 7.13 (b)). We observe that the pendulum is attracted by the plate, inclining toward it. For the time being we will avoid bringing them into contact.

Experiments 7.9 and 7.10 show a new kind of attraction. Up to now we have seen a charged body (conductor or insulator) attracting neutral bodies. In the present cases, on the other hand, the rubbed straw is far away from the pendulum and does not attract it strongly enough to cause the pendulum to incline toward it. But in Section 7.3, we saw that a conducting plate becomes polarized near the rubbed straw. This separation of charges over the body of the plate is shown in Figure 7.13 (b). The plate has no net charge. The sum of its positive charges (close to the rubbed straw) with its negative charges (at the farthest edge) goes to zero. Despite this fact, it attracts a neutral pendulum placed close to its negative side. This is indicated by the inclination of the pendulum in Figure 7.13 (b). As the negative charges of the plate are closer to the disk of the pendulum than the positive charges of the plate, the pendulum is attracted by the plate. That is, the influence of the nearby negative charges upon the disk is greater than the opposite influence of the far away positive charges.

As will be seen in Appendix B, this new kind of attraction was recognized and discovered as a typical electrical phenomenon by Stephen Gray in 1729. Although he discovered this new kind of attraction, he did not know about the polarization of the conductor, nor did he have our modern interpretation of his own experiments. The present day interpretation is due essentially to Aepinus. What needs to be kept in mind, is that although the conducting plate has no net charge, it can attract another neutral body *II* which is close to one of its sides, provided the plate is polarized by a nearby charged body I placed close to the other side of the plate.

Experiment 7.11

In this experiment we again use the thin cardboard rectangle of Experiment 7.6, an electric pendulum and a plastic straw.

We rub the straw in hair and charge the paper disk of the electric pendulum by the ACR mechanism, as described in Section 4.8. When we move the rubbed straw near the charged pendulum we observe a repulsion, as indicated by the inclination of the pendulum relative to the vertical. On the other hand, when the distance between the rubbed straw and the paper disk of the pendulum is greater than or equal to 15 cm, the pendulum remains vertical. Although both of them are negatively charged, the force of repulsion at this distance is too small to be easily detected (Figure 7.14 (a)).

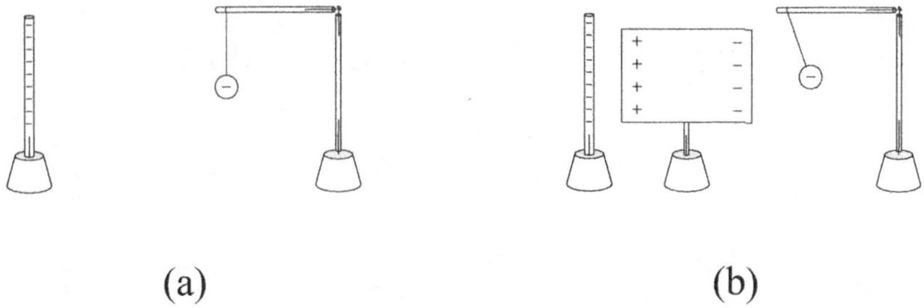

Figure 7.14: (a) A negatively charged straw does not noticeably affect a negatively charged pendulum which is far away from it. (b) When an initially neutral rectangular conductor is placed between them, a repulsion appears.

We now move the rubbed straw vertically near to one edge of the thin cardboard rectangle, preventing them from coming into contact. In the same vertical plane we place the vertical rubbed straw, the thin cardboard rectangle, and the charged pendulum. The charged straw remains close to one vertical edge of the rectangle and the pendulum is initially far away from the other edge. We now slowly bring the pendulum toward the rectangle. We observe that the pendulum is repelled by the rectangle. This repulsion increases when we decrease the distance between them. When there is a distance of 15 cm between the rubbed straw and the vertical projection of the point of support of the silk thread of the pendulum, this thread will be visibly away from the vertical, as it is repelled by the rectangle between the straw and the pendulum (Figure 7.14 (b)).

Now we remove the rectangle, keeping the rubbed straw and the pendulum stationary relative to the ground. The rectangle should be removed along a horizontal direction perpendicular to its plane. After the rectangle has been removed, we observe that the charged pendulum returns to the vertical direction, and the pendulum remains at 15 cm from it.

This experiment is another proof of the polarization of the conducting plate in the presence of the rubbed straw.

Experiment 7.12

It is possible to perform an experiment analogous to Experiment 7.11 utilizing a thin cardboard plate like the one in Experiment 7.6, a rubbed plastic straw and a discharged electric pendulum. The rubbed straw is placed vertically close to one edge of the plate (the left one, for instance), with the discharged pendulum far away from the plate, but in the same vertical plane (Figure 7.15 (a)).

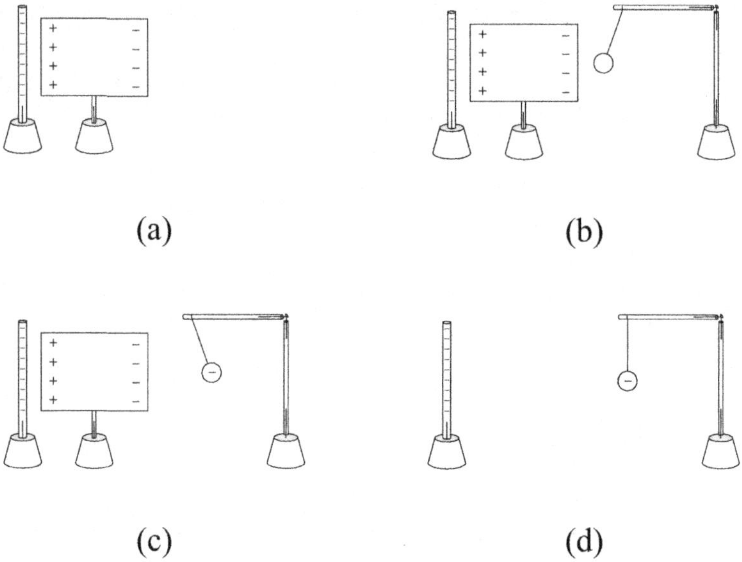

Figure 7.15: (a) Conducting plate close to a charged straw. (b) A nearby neutral pendulum is attracted by the plate. (c) After contact, the pendulum is repelled by the plate. (d) When the plate is removed, the negative pendulum hangs vertically, since it is far away from the negative straw.

We now slowly bring the pendulum toward the left edge of the plate. At a certain distance from the plate, we observe that the paper disk of the pendulum is attracted by the plate, as in Figure 7.15 (b).

When the charged straw is brought even closer to the left edge of the plate, the disk touches the plate and begins to be repelled by it due to the *ACR* mechanism. We then arrive at a situation analogous to Figure 7.14, as in Figure 7.15 (c).

When we remove the plate, the pendulum returns to the vertical (Figure 7.15 (d)). When the rubbed straw is brought near the pendulum, a repulsion appears. This indicates that both bodies (the straw and the pendulum) are electrified with charges of the same sign.

Experiment 7.13

An experiment analogous to Experiment 7.12 involves bringing an initially discharged electric pendulum close to one edge of a discharged conducting plate, without touching it. The pendulum remains vertical. We now slowly bring a rubbed vertical straw near the other edge of the plate. At a certain distance the pendulum begins to be attracted by the plate, touches it, and is then repelled by it. We now remove the plate. When the charged straw is brought near the pendulum, a repulsion appears. This indicates that both of them have charges of the same sign.

7.5 Utilizing Polarization to Charge an Electroscope

Up to now we have seen how to charge a body positively or negatively by friction. Moreover, when we use the *ACR* mechanism, we saw how to electrify a conductor with a charge of the same sign as that as a previously rubbed body. We now utilize the electrical polarization of conductors, together with the fact that the charges move freely on their surfaces, to describe a third mechanism of electrification.

Definitions: Following the definitions presented in Section 7.3, the charging mechanisms described in this Section are called *electrification by induction*, *charge by induction*, *charging by induction*, *electrification by influence*, *charge by influence*, or *charging by influence*.

7.5.1 First Procedure of Electrification by Induction

Experiment 7.14

We build two electroscopes with thin cardboard rectangles of dimensions 10 by 7 cm, A and B, as in Section 6.1. The longer sides will be vertical. There are strips of tissue paper attached to the upper center of these electroscopes. These two electroscopes are placed side by side in a single vertical plane, with their closest sides touching one another, as shown in Figure 7.16 (a). We discharge them by contact with our finger. The two strips remain vertical. We rub a plastic straw in hair so that it receives a good electrification, as indicated by the wall test in Experiment 3.6. This rubbed straw is attached vertically to an appropriate support, and initially far away from the electroscopes.

We slowly bring this rubbed straw near the free vertical edge of electroscope A, without bringing them into contact. It should remain close to this edge, with electroscope A between it and electroscope B. The two strips rise, indicating that each electroscope has become charged as shown in Figure 7.16 (b).

Keeping the rubbed straw close to the free edge of electroscope A, we move electroscope B away from electroscope A. While doing this we should take care to touch only the plastic straw, or its supporting base, holding the plate of electroscope B, without touching its plate or its strip. We observe that the two strips remain raised as shown in Figure 7.16 (c).

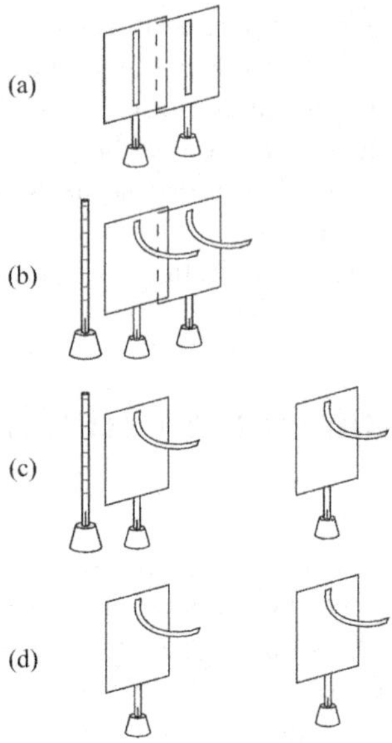

Figure 7.16: First procedure for charging by induction.

After this procedure, we place the rubbed straw far away from both electroscopes. We observe that the two strips remain raised as shown in Figure 7.16 (d), indicating that the electroscopes are charged.

We now hold the rubbed straw horizontally, at the same height as the lower ends of the raised strips of the electroscopes. When we bring the horizontal rubbed straw near the strip of electroscope B, preventing them from coming into contact, we observe a repulsion of the strip. That is, the strip moves away from the rubbed straw, toward the plate of electroscope B. This shows that this electroscope has become electrified with a charge of the same sign as the rubbed straw.

When the rubbed straw is slowly brought near the strip of electroscope A, without coming in contact, an attraction results. That is, this strip moves toward the straw, away from the plate of its electroscope. This indicates that this electroscope has become electrified with a charge of opposite sign to the rubbed straw.

We now remove the rubbed straw. We place both electroscopes in parallel planes, facing one another, with the strips pointing toward one another. By moving these two electroscopes close to one another, but preventing contact between the two strips, we can see their mutual attraction. This again shows that they are oppositely charged (Figure 7.17).

This experiment offers an additional proof of the electrical polarization of conductors. In this case the two electroscopes A and B, which were initially touching one another, behaved as a single conductor. When we moved the

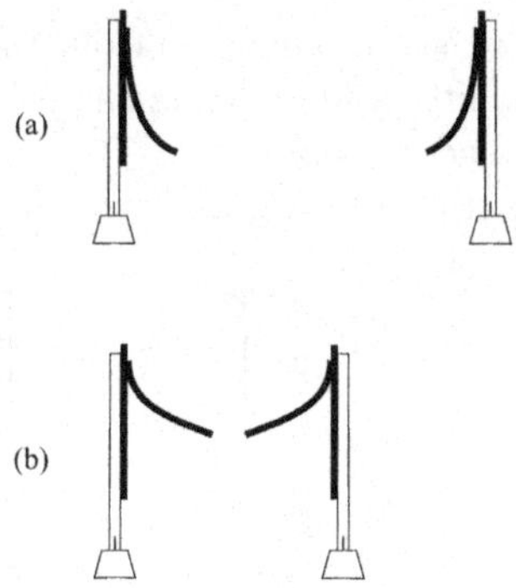

Figure 7.17: Two electroscopes charged by induction become oppositely electrified.

rubbed straw near electroscope A, this electroscope has become electrified with a charge of opposite sign to the straw, while electroscope B has become electrified with a charge of the same sign as that of the straw. When we removed electroscope B, it conserved its charge. The same happened with electroscope A when the straw was removed.

This experiment also shows the conservation of charge. The reason is that the two electroscopes were initially discharged. During the experiments they were insulated from the ground and we did not touch them with the rubbed straw. After they were electrified, one of them received a positive charge and the other a negative charge. This experiment also shows that the electric charges can move freely on the surface of the conductors. It also shows that we can spatially separate positive and negative charges, accumulating each one of them in separate electroscopes.

7.5.2 Second Procedure of Electrification by Induction

Experiment 7.15

We now describe a second procedure to electrify by induction.

The strip on an initially discharged electroscope points vertically downward. We throughly rub a plastic straw in hair to give it a good amount of charge, as in Experiment 3.6. This rubbed straw is attached vertically to an appropriate support, far away from the electroscope.

We slowly bring the rubbed straw near one edge of the electroscope, preventing them from coming into contact. Its strip lifts and remains raised.

While keeping the rubbed straw close to one edge of the electroscope, we touch the other edge with our finger. The strip drops and remains pointing downward.

While keeping the rubbed straw close to one edge of the electroscope, we remove the finger from the other edge. The strip remains pointing downward.

Now we move the rubbed straw far away from the electroscope. After the straw is removed, the strip lifts and remains raised! This indicates that this procedure has charged the electroscope.

These five steps are illustrated in Figure 7.18.

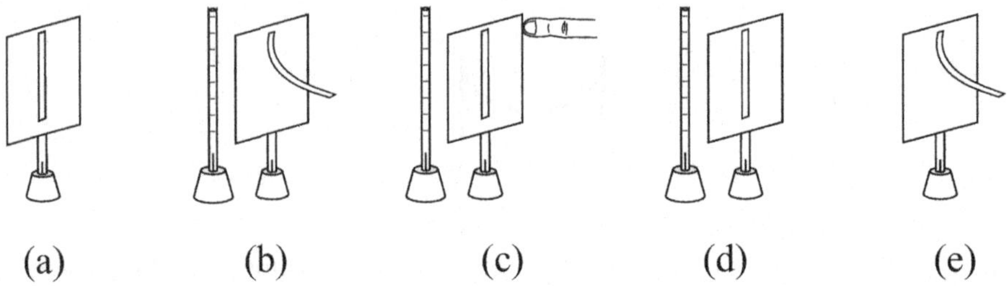

Figure 7.18: Second procedure to charge an electroscope by induction.

The rubbed straw is now placed horizontal at the same height as the lower end of the raised strip. We slowly bring it near this strip, preventing them from coming into contact. We observe that they attract one another, with the strip moving toward it and away from the plate of the electroscope. By moving the rubbed straw upward, we can even cause the free end of the strip to rise above the upper edge of the electroscope (Figure 7.19).

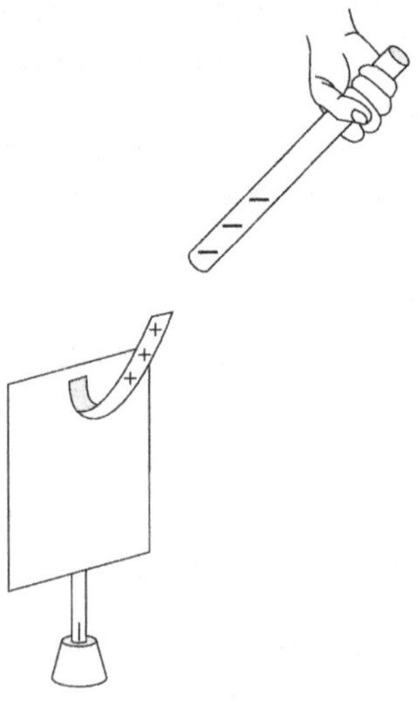

Figure 7.19: Negatively charged straw attracting the strip of the electroscope of Figure 7.18 after the experiment.

This indicates that the electroscope has become electrified with a charge of opposite sign to the rubbed straw.

We can describe what happened using previous results. When the rubbed straw came close to one edge of the electroscope, the electroscope became electrically polarized. The edge close to the straw became electrified with a charge of opposite sign to the straw, while the opposite edge became electrified with a charge of the same sign as that of the straw. When we touched the free opposite edge of the electroscope, we discharged the electricity which had been accumulated on this side. As the rubbed straw remained close to the first edge, the charges at this edge remained opposite to the charge on the straw, due to their mutual attraction. That is, they were not discharged by our finger touching the other edge. When we removed the finger from the second free edge, nothing changed for the charges on the first edge. We then finally removed the charged straw. After this procedure, the charges which were concentrated on the first edge spread out over the whole electroscope. This raised the strip. This distribution of charges is illustrated in Figure 7.20.

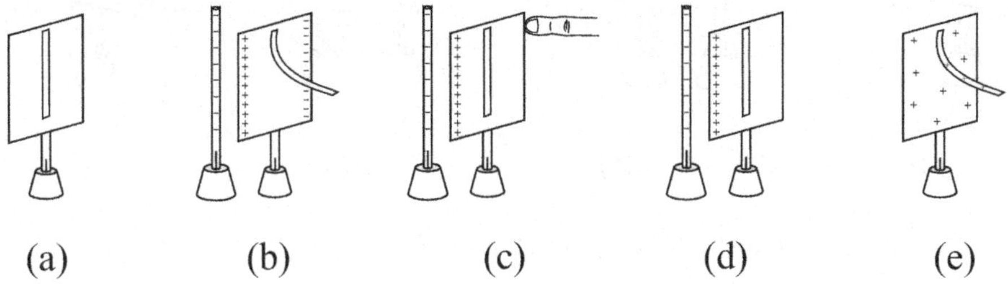

Figure 7.20: Distribution of charges of Figure 7.18.

With this experiment we obtain an opposite effect from Experiments 6.2 and 6.5 where, when we charged an electroscope by friction or by contact, it acquired a charge of the same sign as the body which charged it. In this experiment, on the other hand, the electroscope acquired a charge of opposite sign to the rubbed body which was placed close to it.

In this case the cardboard plate of the electroscope was initially neutral. At the end of the process, it was electrified. In order to obtain this electrification of the electroscope, we used the following procedure. Initially we polarized the electroscope in the presence of a charged body, grounded the free end of the electroscope, removed the contact with the ground, and removed the rubbed body at the end of the process. The grounding was necessary to neutralize the charge which had accumulated at the free end of the conductor due to its polarization. The final result is that the electroscope became electrified. In this case the effect of the grounding was to charge it! This shows that the grounding does not always discharge a body, as was the case in Experiment 4.9.

7.5.3 Third Procedure of Electrification by Induction

Experiment 7.16

Experiment 7.15 can be made in another way utilizing once more an electroscope. Initially we ground one edge of the electroscope by touching it with our finger or connecting it to the ground with a piece of metal wire. While this edge is grounded, we move a rubbed straw near the other side of the electroscope, without touching it. While the rubbed straw is close to this second edge, we remove the grounding from the first edge. We then remove the rubbed straw and observe that at the end of the process this electroscope has become electrified, as indicated by its raised strip (Figure 7.21). When we test the sign of the charge acquired by the electroscope, we see that it is opposite to the charge on the straw.

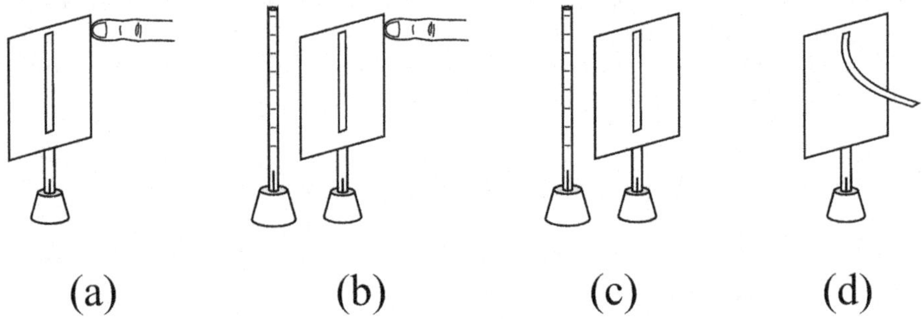

Figure 7.21: Third procedure to charge an electroscope by induction.

The distribution of charges in this experiment is illustrated in Figure 7.22.

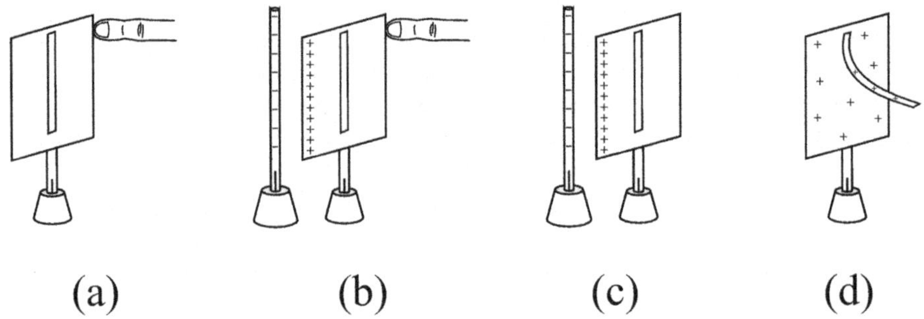

Figure 7.22: Distribution of charges of Figure 7.21.

7.6 The Electric Polarization of Insulators

We will now look at other differences between conductors and insulators.

Experiment 7.17

We now repeat Experiment 7.11 with a neutral Styrofoam plate with dimensions 10 by 7 cm, instead of the thin cardboard plate. Styrofoam is an insulator, while the thin cardboard is a conductor. We rub a plastic straw in hair and it charges a pendulum by the *ACR* mechanism, which then also becomes negatively charged.

When the rubbed straw is at a distance greater than or equal to 15 cm from the charged pendulum, the silk thread remains vertical. Their repulsion is too small to be detected. On the other hand, when this distance is of 15 cm and when the Styrofoam plate is placed between the rubbed straw and the pendulum, we observe that the charged pendulum is repelled by the plate.

This repulsion cannot be directly due to the rubbed straw, as it is at a great distance from the pendulum. This means that this repulsion must be due to a polarization of the Styrofoam. That is, the edge of the Styrofoam closer to the rubbed straw becomes positively charged, while the far away edge becomes negatively charged. What indicates this polarization is the visible repulsion of the negatively charged pendulum when it is at a distance greater than 15 cm from the rubbed straw.

Experiment 7.18

We build now a *plastic electric pendulum*, also called a plastic pendulum (Figure 7.23).

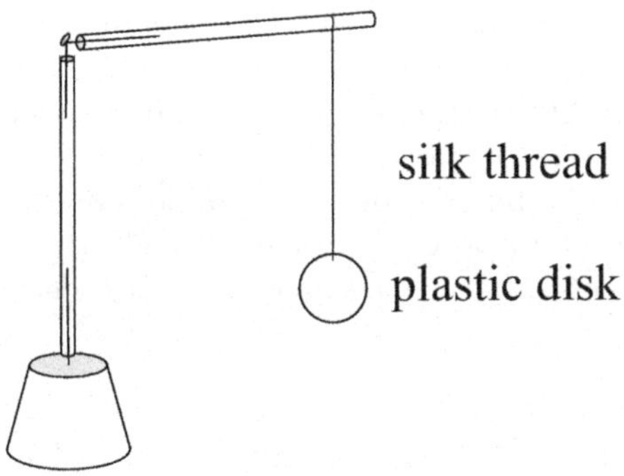

Figure 7.23: Plastic electric pendulum.

In the usual electric pendulum of Section 4.4 we have a paper disk at the lower end of the silk thread. The paper is a conductor. In the plastic electric pendulum we replace the paper disk by an insulator.

Most plastics behave as insulators. However, a few of them behave as conductors. This can be due to the humidity accumulated over their surfaces, or it can be due to their chemical composition. Therefore, initially we should choose a plastic bag which does not discharge an electrified electroscope when the plastic, held in our hand, touches the upper edge of the thin cardboard. This insulating plastic will be used to construct the plastic electric pendulum.

To do this, we cut a disk of 1 or 2 cm in diameter from a thin plastic bag. We make a hole with a needle in this small disk and tie it with the silk or nylon thread.

Before beginning the experiment it is important to verify whether the plastic pendulum is really neutral. In order to know whether it is charged, we move a finger near it and observe whether the plastic disk remains vertically at rest. If this is the case, we say it is neutral. If the plastic disk is attracted by the finger, we say it is charged. We then discard this charged pendulum and build a new neutral one. Often the plastic pendulum may be charged by handling, when we cut or tie the disk.

We rub a plastic straw with hair and bring it close to a neutral plastic pendulum. The plastic disk is attracted by the rubbed straw. But this attraction is much smaller than a similar attraction between a rubbed straw and the paper disk of an ordinary electric pendulum. This force is indicated by a angle of inclination of the disk relative to the vertical when the rubbed straw is at the same distance from both pendulums.

If we allow the rubbed straw and the plastic disk of the plastic pendulum to come into contact, they stick together. That is, the *ACR* phenomenon described in Experiment 4.10 and in Section 4.8 does not happen with a plastic pendulum. The sequence of attraction, contact, and repulsion normally only happens for a conductor. When the body which is being attracted is a neutral insulator, it can touch the attracting body without being repelled by it afterward. The *ACR* mechanism only happens for an insulator after several contacts with the rubbed body, or when we scrape the rubbed body on the insulator.

This is an important difference between conductors and insulators. In order to charge an insulator like plastic, we need to rub it, as in Experiment 2.1. A conductor, on the other hand, can be charged not only by friction, as we saw in Experiments 6.2 and 6.24, but also by the *ACR* mechanism. In this case contact between a conductor and a previously charged body is normally enough for a portion of the charge of the electrified body to be transferred to the conductor.

Experiment 7.19

We repeat Experiment 7.3 with a neutral disk of hard plastic or with a Styrofoam disk, instead of a thin cardboard disk. This time, when we remove the Coulomb proof planes in order to test their charges, we find that none of them has become electrified.

However, we saw in Experiment 7.17 that the Styrofoam becomes electrically polarized in the presence of a rubbed straw. This shows that the polarization which happens in an insulator is different from the polarization of a conductor.

7.7 Does an Electrified Body Attract a Conductor or an Insulator More?

In this Section we discuss an interesting question. First we electrify a plastic straw when we rub it with hair. We place a small, light conductor and a small, light insulator separated from one another on a table. Let us suppose that this conductor and this insulator have the same weight and the same size. When we move the rubbed straw near the conductor and near the insulator, which one will be attracted more? That is, which one will undergo a greater force exerted by the electrified straw?

In Experiments 2.3 and 2.4 we saw that a rubbed plastic attracts conducting substances (like paper and metal) more strongly than insulating substances of the same weight, size and shape (such as plastics or silk).

Experiment 7.20

In this experiment we illustrate the property that a conductor experiences a greater force exerted by a nearby electrified body than an insulator. We will not need to weigh the conductor or the insulator. To do this, we use two plastic pendulums of the same size and shape, made of the same materials (Figure 7.23). In pendulum II a paper disk or an aluminum foil disk is attached to the plastic disk. Due to the extra material attached to it, it weighs more than pendulum I to which nothing has been added.

Before beginning the experiment, we bring a finger near both pendulums. When they are not attracted by the finger, this indicates that they have not been electrified during their construction. This is not always to achieve as the plastic pendulum may easily acquire a net charge during its construction (friction with our hand while cutting or tying the plastic disk to the silk thread). If this happens, there is a simple procedure to discharge the pendulum. We only need to wait a long time (several hours), until the plastic disk looses this charge due to the surrounding air. In Section 7.14 we discuss this topic in greater detail.

From now on we will assume that both pendulums are neutral.

We now bring a rubbed straw close to both pendulums, always preventing the straw and the disks of the pendulums from coming into contact. The plastic pendulum is slightly attracted by the rubbed straw (Figure 7.24 (a)). The pendulum with the paper disk, on the other hand, is much more strongly attracted than the plastic pendulum (Figure 7.24 (b)). This force is indicated by the angle of inclination of each pendulum to the vertical (supposing the electrified straw at the same distance from the disk of the attracted pendulum). Although pendulum II is heavier than pendulum I, it experiences a greater attractive force than the first pendulum.

Experiment 7.21

An analogous experiment can be performed with two pendulums on which we replace the plastic disks with small Styrofoam balls. Like plastic, Styrofoam is also an insulating material. We place spheres of the same size on both pendulums. After this procedure, we cover pendulum II with aluminum foil. We then move a rubbed straw near both pendulums. We observe that the pendulum with aluminum foil is attracted more than pendulum I, to which nothing has been added.

Experiment 7.22

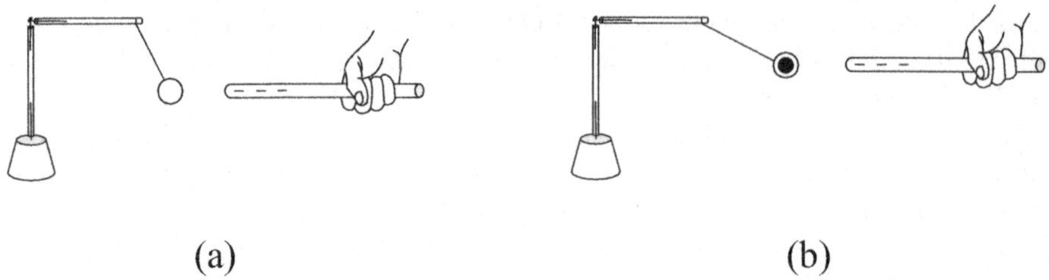

Figure 7.24: (a) A plastic pendulum is weakly attracted by a rubbed straw. (b) A plastic pendulum to which a conducting disk has been added, on the other hand, is strongly attracted by the rubbed straw, despite its greater weight.

We now construct two "insulating pendulous threads," as in Figure 7.25. They are analogous to Gray's pendulous thread (Figure 4.28). But now we replace the wood skewer by a plastic straw, and the cotton thread by a flexible plastic strip. Both insulating pendulous threads should have the same length and shape, and be made of the same material. We now wrap the plastic strip of pendulum *II* with a light cotton thread, like an helix around it. When everything is ready, we test both pendulous threads in order to check that they are neutral before beginning the experiment.

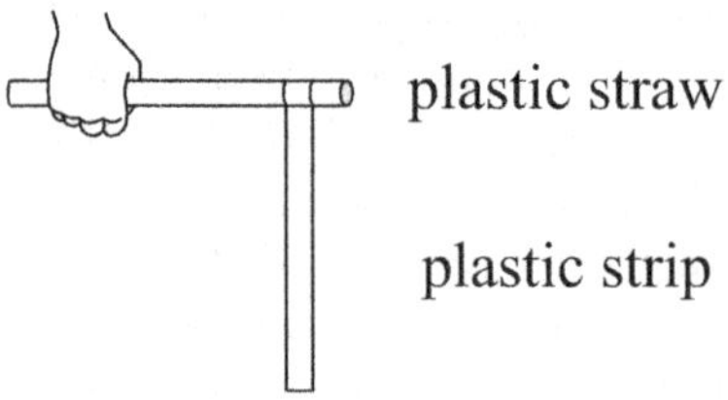

Figure 7.25: An insulating pendulous thread.

We move a rubbed straw near both pendulous threads. We observe that pendulous thread *II* with the conducting cotton thread is attracted more than pendulous thread *I* to which nothing has been added (Figure 7.26). Although the pendulous thread *I* is lighter than the pendulous thread *II*, it is less strongly attracted than this second pendulous thread.

These experiments show that a conductor experiences a greater force than an insulator, with both forces being exerted by the same electrified body. Du Fay and Aepinus are among the researchers who discovered this fact experimentally.[6]

[6][DF33d, pp. 233-234] and [Aep79, pp. 261, 274, and 309-315].

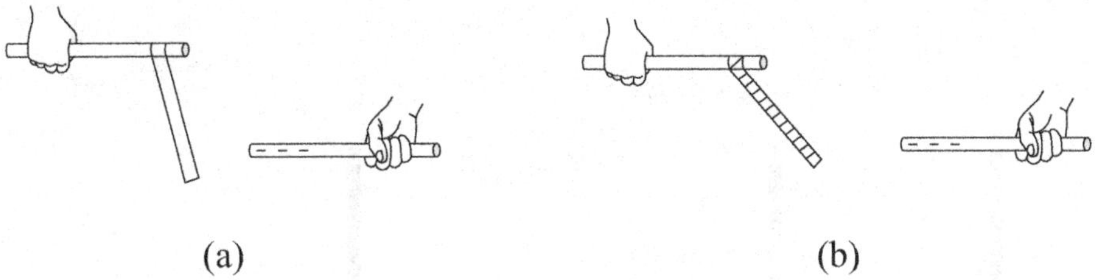

Figure 7.26: (a) An insulating pendulous thread is less attracted by an electrified body than another insulating pendulous thread to which a conducting thread has been added (b).

7.7.1 Discussion of Gray's Electric Pendulum

As mentioned in Section 4.6, in 1720 Gray described an electric pendulum in which he tied a feather to a silk thread connected to a stick. At that time no one knew the distinction between conductors and insulators. This means that his use of a silk thread was coincidental. Silk is an insulator material. At that time Gray might just as well have employed a linen or cotton thread, which are conductors. Gray heated a piece of brown paper (transforming this paper into an insulator) and charged it by friction. By bringing this fine brown paper near the feather connected to the silk thread, he could raise it until the silk thread became horizontal, that is, at right angles to the vertical stick. By lifting the rubbed brown paper even higher, he could make the silk thread extend itself above the horizontal plane, remaining upright in the air, as in Figure 4.21 (c). In the sequel to this experiment he mentioned the following:[7]

> I then repeated this experiment without the feather, viz. by a single thread of silk only of about 5 or 6 inches long [13 or 15 cm], which was made to stand extended upright as above-mentioned, without touching the [rubbed and heated] paper; [...]

An illustration of this experiment appears in Figure 7.27.

This shows that the feather had only a secondary role in this experiment, as he could lift the silk thread even without the feather. There are two possible explanations for this curious effect. The first is that when he tied and untied the feather in the silk thread, the silk behaved as an insulator and became charged on its lower end by friction with his hands. The heated and rubbed brown paper may had been electrified with a charge of opposite sign to the charged silk. When Gray brought these two substances together, they attracted one another. Gray could then lift the silk thread above the horizontal. The second possibility is that when he tied and untied the feather, the silk thread may have become humid, perhaps due to sweat from Gray's hands. If this was the case, the silk thread would then behave as a conductor. When he moved the heated and rubbed brown paper near the thread, the conducting thread would become

[7][Grab, p. 107].

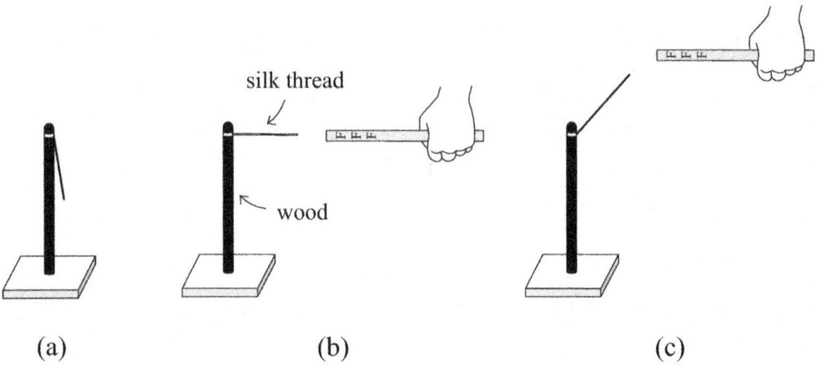

Figure 7.27: Gray attracting a silk thread by bringing a piece of rubbed paper near it.

polarized. That is, its lowest end, which was closer to the charged paper, would acquire a charge of opposite sign. Due to the strong attraction between the charges of the paper and the charges located at the free end of the thread, Gray would have been able to lift the thread above the horizontal. This would be analogous to our Experiment 6.6.

If the silk thread were dry and discharged, it would behave as a usual neutral insulator. In this case Gray would not be able to lift it above the horizontal, even when he brought the charged brown paper near the thread. Normally the attraction between a charged body and a neutral insulator is much weaker than the attraction between a charged body and a neutral conductor. It is also much weaker than the attraction between two oppositely charged bodies.

7.8 Forces of Non-Electrostatic Origin

As we saw in Chapter 5, two positively charged bodies repel one another, two negatively charged bodies repel one another, while a positive body and a negative body attract one another. Sometimes this is expressed by saying that charges of the same sign repel one another, while opposite charges attract. The electric forces which these bodies exert one upon the other when they are at rest are called electrostatic forces.

In all situations in which there are two or more charges at rest relative to one another in a configuration of stable equilibrium, non-electrostatic forces are necessary in order to balance the electrostatic forces exerted between the interacting charges.[8] We illustrate this here in the case of a negatively charged conducting sphere. The sphere may have been electrified, for instance, by a plastic straw rubbed in hair. The charges on the conductor repel one another. After reaching equilibrium, they are distributed uniformly over the surface of the sphere (Figure 7.28).

[8]See [AH07, Section 5.3], [AH09, Section 5.3], and the references therein.

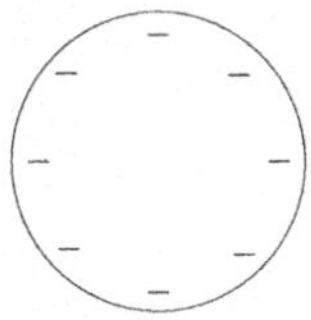

Figure 7.28: A uniformly charged sphere.

Let us consider the negative charge at the top of the sphere. It is repelled by all other negative charges. Therefore, it is acted upon by a vertical electrostatic force pointing away from the center of the sphere. A force of non-electrostatic origin pointing downward, having the same intensity as the upward pointing electrostatic force, is needed to keep this negative charge at rest. In this specific situation, this non-electrostatic force is sometimes called a contact force. But its origin is not clearly understood. We also don't know how it is produced, etc.

The same situation happens when the conducting sphere is positively charged. And the same reasoning is valid for an insulating sphere that is uniformly electrified, either negatively or positively.

7.9 Microscopic Models of Conductors and Insulators

In order to understand this whole set of experiments showing the differences and similarities between conductors and insulators, microscopic models of these bodies are made. These models help us comprehend and visualize the processes being described here. The models have been created utilizing the results of experiments analogous to the ones described in this book. Afterward the procedure is inverted. That is, these models are postulated and then used in order to illustrate or describe what is happening in the experiments.

This variant behaviour leads to two different microscopic models for polarized conductors and polarized insulators. For conductors we suppose the existence of free charges. We assume that in conductors there are charges which are not attached to the molecules of the material, and are thus free to move throughout the conductor. When the conductor is neutral, these free charges experience no net macroscopic movement and do not generate external effects. On the other hand, when we move a charged body near this conductor, it gets polarized. In particular, the portion of the conductor which is closest to the charged body becomes electrified with a net charge of opposite sign, while the opposite portion of the conductor becomes electrified with charges of the same sign as those of the charged body (Figure 7.29). These polarized free charges can be transferred to other conductors if these other conductors come into contact with this polarized conductor.

We assume the existence of a force of non-electrostatic origin in order to prevent these free charges from leaving the surface of the conductor, except in

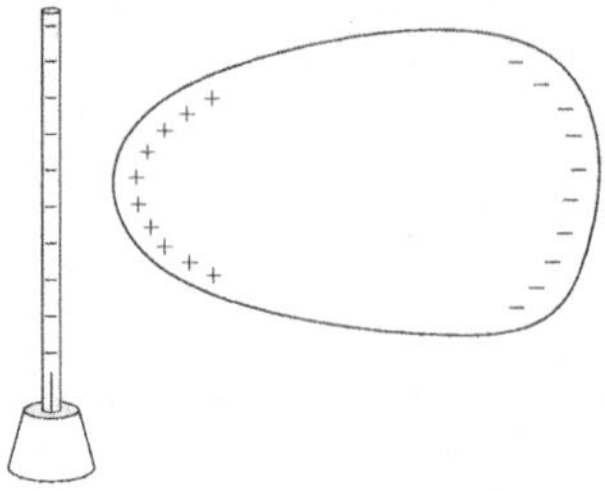

Figure 7.29: Microscopic model of an idealized polarized conductor close to another charged body.

breakdown conditions for which there are electric discharges through the air.

The grounding of a conductor in the presence of a nearby charged body, as shown in Experiments 7.15 and 7.16, is illustrated in Figure 7.30.

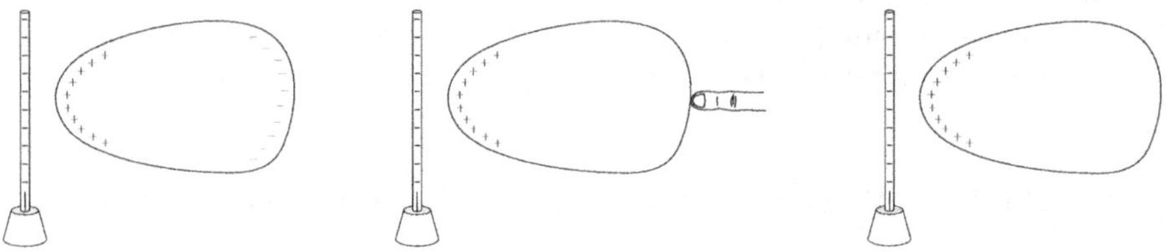

(a) (b) (c)

Figure 7.30: Grounding of a conductor in the presence of a nearby charged body.

We now present the microscopic model for an insulator. In this case we suppose that when we move a charged body near an insulator, only the molecules of the insulator become polarized. That is, the portions of the molecules of the insulator which are closer to the charged body become electrified with charges of opposite sign to the charged body. The portions of the molecules of the insulator which are farthest from the charged body, on the other hand, become electrified with charges of the same sign as this body. For insulators there would be no net motion of free charges, only a polarization of its molecules (Figure 7.31 (a)).

In the interior of the insulator there would be equal quantities of positive and negative charges, very close to one another. Considering any small volume inside the insulator containing many molecules, the net sum of these charges would be approximately zero. This means that we can consider the interior of the polarized insulator as macroscopically neutral. But this will not be the case for the surface. The net effect of these molecular polarizations would be that the surface of the insulator closest to the charged body would behave as if it were electrified with charges of opposite sign to the external charged body. The surface of the insulator which is farthest from the charged body, on the other hand, would behave as if it were electrified with charges of the same sign as this

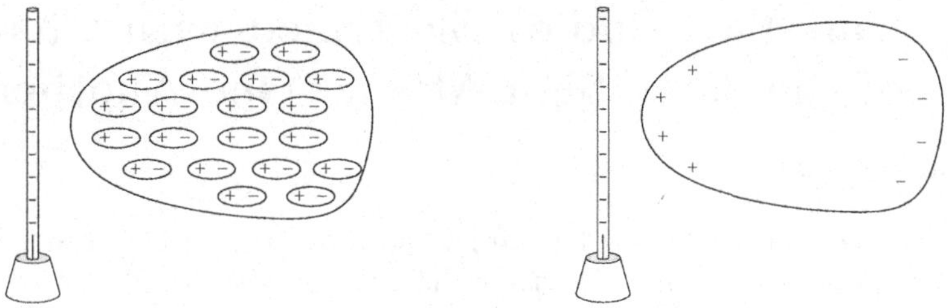

(a) (b)

Figure 7.31: (a) Microscopic model of an idealized insulator polarized in the presence of another charged body. (b) Effective polarization of the insulator in situation (a).

body. This effective polarization is illustrated in Figure 7.31 (b).

Once more, a force of non-electrostatic origin is required in order to prevent the polarized charges in each molecule from moving through the insulator.

The polarization presented in Figure 7.29 is greater or more intense than the polarization presented in Figure 7.31 (b). These Figures have been made deliberately. We are assuming conductors and insulators of the same shape and size, both at the same distance from a straw charged with the same intensity in both cases. The reason for the amounts of charge in these Figures was given in Section 7.7. Experiments show that the force exerted by a charged body upon a conductor is greater than the force exerted by this charged body upon an insulator. This indicates that the polarization of a conductor is greater than the effective polarization of an insulator. The intensity or degree of this polarization is represented by the number of opposite charges distributed over both sides of the polarized body. For a conductor there are more polarized charges than for an insulator, as represented in Figures 7.29 and 7.31 (b).

In addition, we increase the number of polarized charges in conductors and insulators, by decreasing their distance to the nearby charged body which is inducing these charges. We also increase the number of polarized charges by increasing the electrification of the nearby body which is polarizing the conductor and insulator.

In reality, no body is a perfect conductor or a perfect insulator. As a result, these microscopic models are idealizations. Real bodies present characteristics of both behaviours, to a greater or lesser extent. There is a gradation between good conductors and good insulators.

In any event, these idealized models are extremely helpful in order to help us understand and visualize what happens in many electrical phenomena.

7.10 Can Two Bodies Electrified with Charges of the Same Sign Attract One Another?

Experiment 7.23

We repeat Experiment 6.5 (Figure 6.7). But now the negative straw is moved even closer to the strip of the negatively charged electroscope. We observe that for distances smaller or equal to a certain value, of the order of 2 to 4 cm, the strip is no longer repelled by the straw, being attracted by it. The strip touches the straw and remains attached to it.

Experiment 7.24

We repeat Experiments 4.7 and 5.23 (Figures 4.18 and 5.27). But now the negative straw is moved even closer to the negative disk of the pendulum. We observe that the inclination of the pendulum from the vertical increases when the distance between the straw and the pendulum goes from 15 to 5 cm, approximately. This shows that the intensity of the repulsive force increases when the distance between them decreases in this range.

However, for distances less than or equal to certain value, of the order of 5 cm, there is no longer a repulsion between the negative straw and the negative disk. At these small distances they attract one another. The disk touches the negative straw a second time and is again repelled by it.

After 2 or 3 of these contacts between the negative straw and the disk of the pendulum, the same phenomena are once more observed, but over somewhat different distances. When the disk has a greater degree of electrification, the repulsion between the disk and the negative straw can be observed at a distance larger than before, of the order of some 20 cm. The intensity of the repulsive force increases when this distance is decreased between 20 cm and a lower limit of 2 or 3 cm. When the distance between the straw and the disk has a value less than or equal to this lower limit, there is again an attraction between them, and the *ACR* mechanism comes into play.

These experiments can be understood on the basis of the principles discovered so far.

Let us suppose that a body *I*, an insulator, has been charged negatively by friction. It is close to a neutral body *II*, a conductor, which has no net charge. There will be an attractive force between them. This attractive force is due to the polarization of body *II* in the presence of body *I*, as in Figure 7.29. We represent their attractive force by $F_A > 0$ (Figure 7.32 (a)). We now electrify body *II* with a charge of the same sign as the charge of the attracting body *I*. This can be done, for instance, by the *ACR* mechanism. This will generate a new force between them. It is repulsive and it will be represented here by $F_R < 0$. In Figure 7.32 (b) we present this new force of repulsion, disregarding the previous attractive force due to the polarization of the conductor. The new negative charge on the conductor is represented in the middle of body *II* only to distinguish it from the polarized charges. This new charge on body II will tend to polarize body I, generating an attractive force between them, but we will disregard this small component here. In any event, the previous attraction which occurred between them will still remain. The net force will be given approximately by $F_N = F_A + F_R$. If $F_A > |F_R|$, the net force will be attractive. If $F_A < |F_R|$, the net force will be repulsive. In Figure 7.32 (c) we show an example for which $F_A < |F_R|$.

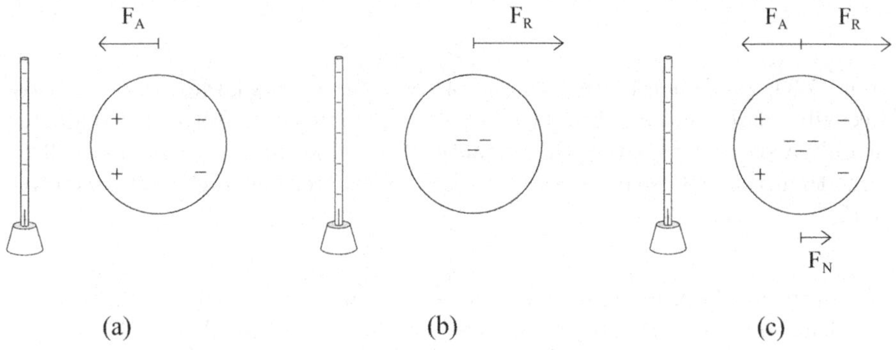

Figure 7.32: (a) Attractive force due to the polarization of a neutral conductor in the neighborhood of an electrified insulator. (b) Ideal repulsive force between a negative insulator and a negative conductor, assuming that the conductor was not polarized by the negative straw. (c) Net force $F_N = F_A + F_R$ between an electrified insulator and a conductor, assuming the conductor is electrified and polarized. Situation for which $F_A < |F_R|$.

From what we have seen up to now, we can give three conditions for which we can have a net attractive force between these two bodies electrified with charges of the same sign.

- The original force of attraction is independent of the value of the new charge given to the initially neutral body II, while the new force of repulsion depends upon the value of this new charge. If this new charge is very large, $|F_R|$ will usually be greater than F_A (Figure 7.33 (a)). By decreasing the magnitude of this new charge on body II, we can decrease the magnitude of the new repulsive force in such a way that a net attraction will remain between the two bodies having a net charge of the same sign (Figure 7.33 (b)).

- Suppose the conductor and the insulator are both negative and that $|F_R| > F_A$, in such a way that there is a net repulsion between them (Figure 7.34 (a)). When we increase the amount of charge in the insulator, we increase the intensity of the repulsive force $|F_R|$. The intensity of the attractive force F_A increases, but it increases faster than the increase in $|F_R|$. The reason is that we also increase the number of polarized charges upon the conductor, as see in Section 7.3 (Figure 7.10). As an example, if the charge

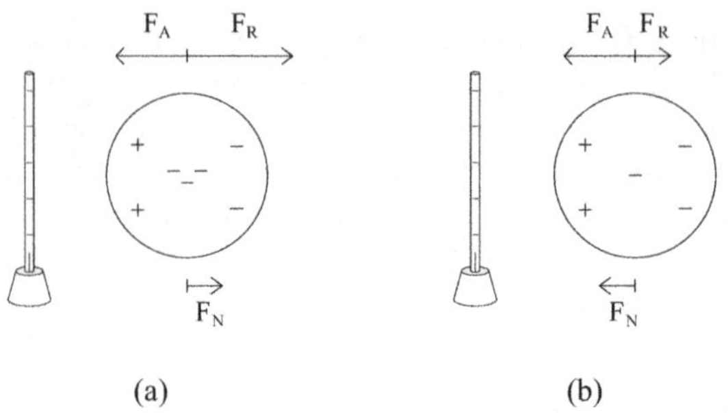

Figure 7.33: (a) When the net charge of the conductor is large, a net repulsive force will remain between it and the negative insulator. (b) When the conductor has only a small net charge, the attractive force due to its polarization will be larger than the repulsive force due to its net charge, yielding a net attractive force.

of the insulator increases three times, $|F_R|$ also increases about three times. But on the other hand, F_A increases about nine times. When the amount of charge on the insulator is increased, there will be a point at which F_A will be larger than $|F_R|$, yielding a net attractive force between them (Figure 7.34 (b)).

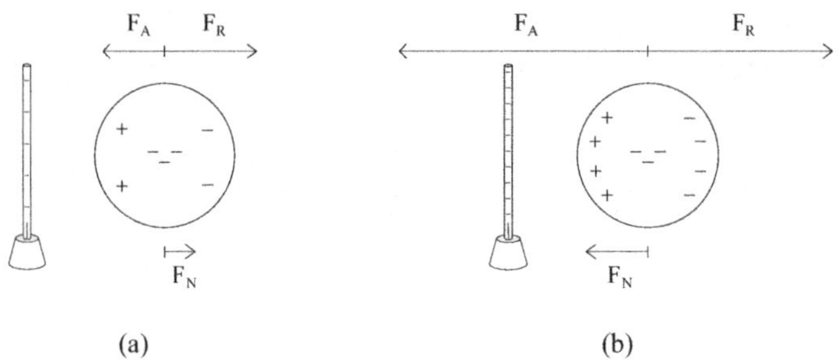

Figure 7.34: (a) Repulsive force larger than attractive force. (b) F_A increases more than $|F_R|$ due to an increase in the amount of charge in the nearby insulator. Here we present a situation for which the attractive force has become larger than the repulsive force.

- There is also another situation which can produce a net attractive force between two bodies having charges of the same sign. Suppose the conductor and the insulator are both negative and that $|F_R| > F_A$, in such a way that there is a net repulsion between them when they are separated by a distance d (Figure 7.35 (a)). The intensities of these two forces of attraction and repulsion behave differently depending on the distance between the two bodies. By decreasing their distance, we increase the magnitude of F_R due only to the approach between the negative charge of I and the net negative charge of II. The attractive force F_A, on the other hand, increases for two different reasons. (a) The first is due to the reduced distance between the negative charge of I and the polarized charges of II. (b) The second reason is that the number of polarized charges on II also increases when the distance between I and II decreases, as see in Section 7.3 (Figure 7.8). As the force depends not only upon the distance but also upon the number of charges in the bodies, this more intense polarization will produce, accordingly, a larger attractive force. This means that if bodies I and II are very close to one another, there may remain a net attraction between them even when both of them have a net charge of the same sign (Figure 7.35 (b)).

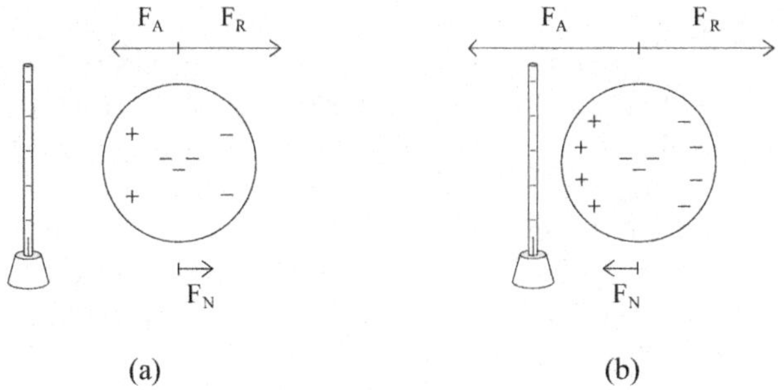

Figure 7.35: (a) Repulsive force $|F_R|$ larger than the attractive force F_A. (b) When the distance between the two bodies decreases, F_A increases more than $|F_R|$. Here we present a short distance for which the attractive force becomes larger than the repulsive force.

The sign of the net force (that is, if it is attractive or repulsive) will depend on the values of the two charges, the distance between the bodies, their shapes, their sizes and their intrinsic properties (that is, if they are conductors or insulators, as this will affect their degrees of polarization).

Du Fay himself knew that in certain circumstances two bodies electrified with charges of the same sign could attract one another:[9]

> In order to succeed in these experiments [of attraction between oppositely charge bodies, and repulsion between electrified bodies having charges of the same type], 'tis requisite that the two bodies, which are put near one another, to find out the nature of their electricity, be rendered as electrical as possible; for if one of them was not at all, or but weakly electrical, it would be attracted by the other, though it be of that sort, that [if well rubbed] should naturally be repelled by it. But the experiment will always succeed perfectly well, if both bodies are sufficiently electrical.

[9][DF, p. 265].

Aepinus gave a first draft of a mathematical explanation of possible attractions between two bodies carrying charges of the same sign. He showed theoretically and experimentally that if two bodies electrified with charges of the same sign are brought sufficiently close together, or if one of the two charges involved is very much weaker than the other, the effects of mutual polarization may be sufficient to change the normal repulsive force into an attraction.[10]

We will not go into details here, but a modern mathematical analysis indicating the conditions for which electrified bodies, having charges of the same sign, can attract one another, can be found, for instance, in Maxwell's work, in a paper by Melehy, and in Jackson's book.[11]

7.11 The Conductivity of Water

In Section 6.3 we saw that fresh water discharges an electrified electroscope. This means that it behaves as a conductor for the usual electrostatic experiments described in this book. On the other hand, when it is subjected to a potential difference of a few volts to a few hundred volts, it behaves as an insulator, as seen in Section 6.6. In the experiments of Section 2.5 we were dealing with a potential difference of a few thousand volts, when the water behaves as a conductor. There are reasons for this behaviour of water. One reason is that in its natural state fresh water contains positive ions, H_3O^+, and negative ions, OH^-, in addition to molecules of H_2O. Moreover, fresh water contains many salts, minerals, and impurities which abound in electrically charged particles, also called ions. In the presence of high potential differences, these electrified ions move in water, giving it its conducting behaviour.

We now analyze the experiments of Section 2.5. Let us suppose that the water drop of Gilbert's experiment, Experiment 2.10, is over a conducting surface, such as wood, metals, or most solids. When we bring a rubbed plastic straw near it, the drop changes its shape. That is, it deforms pointing toward the rubbed section of the straw. If there is low friction between the water and the surface on which it is resting, the drop can even move as a whole toward the straw. With the results seen in this Chapter, it is possible to illustrate what is happening in this experiment. This is done in Figure 7.36. Let us suppose that the straw was rubbed with hair, acquiring a negative charge. When it is brought near the drop, the water initially polarizes as a whole, analogous to what we saw in Figure 7.29. That is, it becomes positively charged in the region closer to the rubbed straw and negatively charged in the farthest region which is in contact with the dry surface. As we have assumed a conducting solid surface, there is an electrical neutralization over the section of the water which is in contact with the surface. This is analogous to the grounding in Figures 7.20, 7.22, and 7.30. Consequently, the water ends up becoming electrified as a whole, i.e., with a net positive charge, opposite to the sign of the charges on the straw. As charges

[10] [Aep79, pp. 126 and 315-325], [BWc], and [Hei99, pp. 396-398].

[11] [Max81, Chapter VII: Theory of electrical images, pp. 80-88], [Mel98], and [Jac99, Section 2.3].

of opposite sign attract one another, the drop deforms and points toward the rubbed straw. It can even move as a whole toward it.

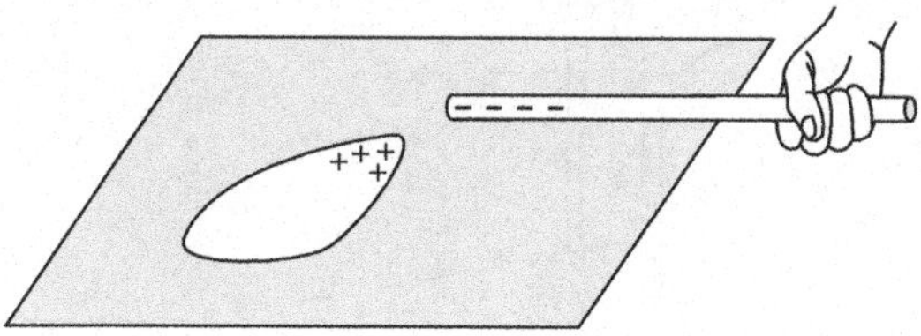

Figure 7.36: Electric charges on a water drop close to a rubbed plastic straw.

That is, we believe that this behaviour of the water drop is not due to a simple orientation or organization of polarized molecules of water (although an H_2O molecule has no net charge, it is naturally polarized, like a permanent electric dipole). If there were only the organization or alignment of the polar molecules in water, due to the presence of the nearby rubbed straw, we would have something analogous to Figure 7.31 (a). Instead of this, it seems more reasonable to say that the water drop as a whole, supported upon a conducting surface and being close to a rubbed straw, acquired a net charge. Moreover, this net charge upon the drop should be of opposite sign to the rubbed straw, as represented in Figure 7.30 (c).

Something analogous happens in Desaguliers's experiment (Experiment 2.8). We have a stream of fresh water falling from a tap. For the high potential differences of this experiment, this water stream behaves as a conductor. Moreover, supposing a continuous stream of water, it is grounded by contact with the metal tap, which is in contact with the ground. Let us suppose that we move a negatively charged plastic straw near the water stream. The water stream is initially polarized (due to the motion and separation of the ions present in water, like H_3O^+, OH^-, etc.), analogous to Figure 7.29. The section of the stream closer to the negative straw becomes positively charged. On the other hand, the negative charges in the farthest section of the stream are neutralized due to the grounding of the water stream. This is analogous to the grounding of Figures 7.20, 7.22, and 7.30. For this reason, the stream as a whole should become positively charged, mainly in the section closer to the negative straw. There is then an attraction between the negative charges of the plastic straw and the positive charges in the stream. Consequently, the stream as a whole bends toward the straw. Figure 7.37 illustrates the distribution of charges along the water stream.

7.12 Is it Possible to Electrify Water?

Water behaves as a conductor in the usual experiments of electrostatics. Therefore, it is possible to electrify it as is usually done with solid conductors. To do this, it must be kept on an insulating receptacle. In this way we prevent its discharge to the ground.

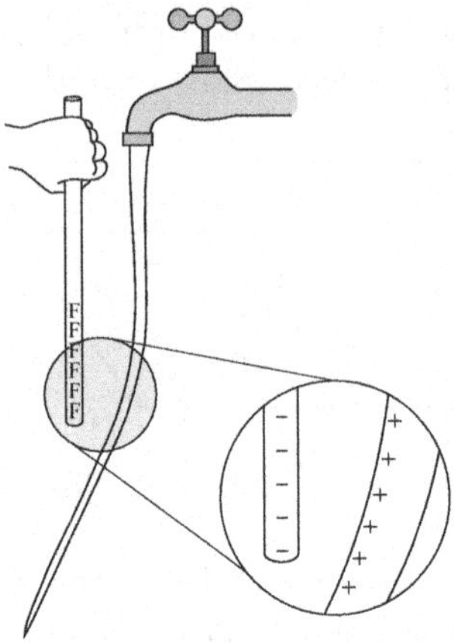

Figure 7.37: Electric charges on a water stream close to a rubbed plastic straw.

As discussed in Section 7.11, in Gilbert's experiment a small amount of water was attracted toward a rubbed amber. Probably the water as a whole was electrified in this situation. But Gilbert did not analyze if the water had been electrified in this case, he only observe its attraction. Perhaps Gray in 1731 was the first to electrify water and to confirm its electrification.[12] He placed water on insulating support made of resin or glass. He then brought an electrified tube 3 or 4 times near the water. After this procedure, he moved the electrified tube far away from the water. By approaching a pendulous thread, he observed it being attracted by the water. This proved that the water was electrified by the electrified tube when the tube was close to the water. The charging mechanism in this case was probably minute sparks between the tube and the water. This caused a charge transfer from the tube to the conducting water. The water could store this acquired charge due to the fact that the water was kept on an insulating support. This experiment was later on confirmed by Du Fay.[13]

With appropriate adaptations, it is possible to reproduce many experiments of Chapter 7 utilizing water inside a plastic receptacle, instead of using a thin cardboard attached to a plastic straw. The conducting behaviour of the cardboard will be assumed by the water. The plastic receptacle will avoid the electrical grounding of the water, just like the plastic straw insulated the cardboard of the electroscope.

7.12.1 Kelvin's Electrostatic Generator

One of the most fascinating experiments showing that water behaves as a conductor in the usual situations of electrostatics was conducted in 1867 by W. Thomson (Lord Kelvin) (Figure 7.38).

[12] [Grad] and [Hei99, p. 253].
[13] [DF33a, p. 34] and [DF33c, p. 84].

Figure 7.38: W. Thomson (Lord Kelvin) (1824-1907).

He built an instrument which is known as water dropper, water-dropping electrical machine, water-drop generator, or Kelvin's electrostatic generator.[14] We present here the main aspects of the generator. Initially water is placed on an insulating receptacle, like a plastic cup. In the bottom of this cup there is a hole or dripper. In the beginning of the experiment, they remain closed. We connect a metal ring to the cup with an insulating material. The ring should be very close to the hole or dripper, at a distance of a few centimeters from it. We rub a plastic straw in hair to charge it negatively. The negative straw is then scratched over the metal ring in order to charge it negatively. After this procedure, the plastic straw is moved away from the ring. Because water behaves as a conductor, it becomes polarized due to the presence of the nearby negative ring just below it. The lower portion of the water becomes positively charged, while the upper portion which is in contact with air becomes negatively charged (Figure 7.39 (a)). This is analogous to the polarization described in Figure 7.29.

After the ring has been negatively charged, we open the hole or dripper. The water droplets should pass through the center of the ring, without coming into contact with it. The lower portion of water in the upper receptacle is positively charged. Therefore, the same will happen with the droplets. These positive droplets are collected inside a second insulating receptacle placed below the ring (Figure 7.39 (b)). While the water droplets continue to fall, the amount of charges accumulated in the lower receptacle increases. This Figure shows the main aspects of the working mechanism of Kelvin's generator.

Normally Kelvin's generator works with two drippers. One of the rings is positively electrified and the other ring is negatively electrified (Figure 7.40). A metal wire with a metal sphere in its upper end is connected inside each lower receptacle. The two metal spheres should be very close to one another, with a typical distance of a few millimeters. The water droplets falling through the negative ring are positively charged, while those falling through the positive ring are negatively charged.

As time goes on, the amount of charges accumulated in each lower receptacle

[14] [Tho], [Llo80], and [CA08].

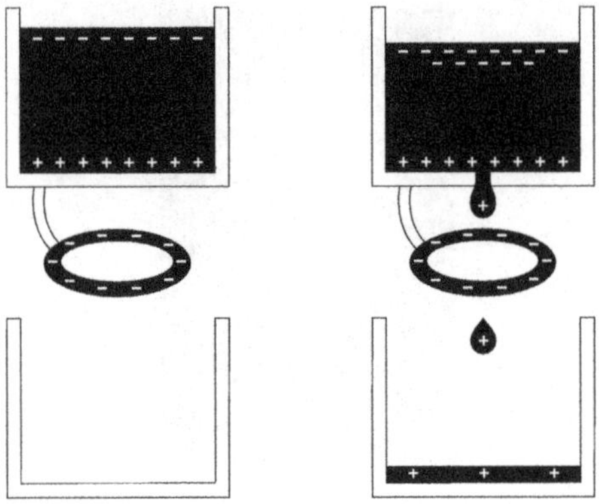

(a) (b)

Figure 7.39: (a) Water polarization due to the nearby negative ring. (b) The water droplets fall electrified with a positive charge.

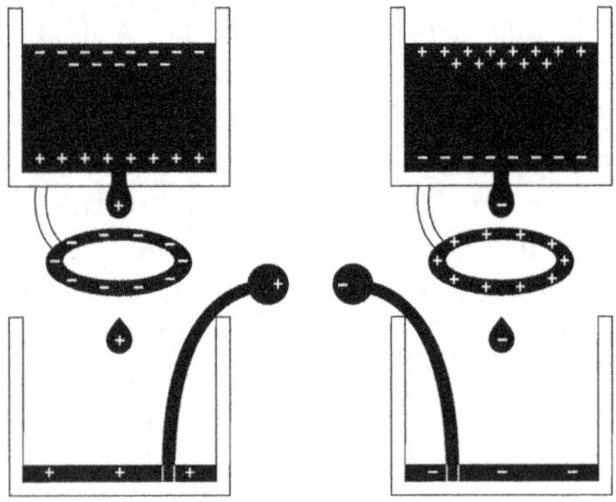

Figure 7.40: First phase of the operating process of Kelvin's generator.

is increased. Therefore, the potential difference between the two metal spheres also increases. Dry air is normally a good insulator, especially if under an electric force per unit charge below a certain limit. When the force per unit charge is higher than this limit, it becomes a conductor and an electric discharge happens through the air, a spark. This limit is called breakdown electric field or corona breakdown. At atmospheric pressure it values approximately 3×10^6 V/m. When the electric force per unit charge in the region between the two spheres goes beyond this limit, there is an electric discharge through the air (Figure 7.41).

With this discharge, there is a neutralization of the opposite charges which were accumulated in the lower receptacles. If the dripping continues after the spark, the process goes on as in Figures 7.40 and 7.41. The time interval between two consecutive sparks will depend upon the dripping frequency, the distance

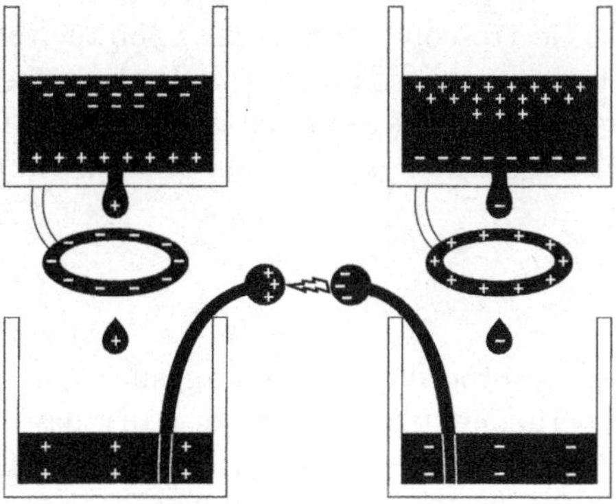

Figure 7.41: Electric discharge between the oppositely charged spheres.

between the metal spheres, the amount of charge in the rings and the distance between the rings and the drippers. A typical order of magnitude is a spark for each 10 seconds. The sparks will continue while the dripping goes on.

In this experiment two important things are shown. The first is that water behaves as a conductor in the usual experiments of electrostatics, as we saw in Sections 6.3 and 7.11. The second one is that if there is an electrified body close to where water is dripping, the water droplets will become electrically charged, with a net charge have an opposite sign to the sign of the charges in the nearby electrified body.

7.13 The Conductivity of Air

Experiment 7.25

It is easy to electrify an electroscope in a dry day. We only need to rub a plastic straw in hair and then scratch this straw over the thin cardboard of the electroscope, as we saw in Experiment 6.2 (Figure 6.5). The electroscope remains charged for several seconds or for a few minutes after this procedure. This means that dry air is a good insulator.

However, it is not a perfect insulator. After several minutes the electroscope is totally discharged, as we saw in Experiment 6.21. In any event, it can be considered as a good insulator, according to the definition given in Subsection 6.7.1.

Experiment 7.26

Experiment 7.25 is repeated in a humid and rainy day. The strip of the electroscope remains raised while we scratch the rubbed straw in the cardboard of the electroscope, as in Figure 6.5 (b).

We then remove the rubbed straw. The strip drops soon afterward, returning to the situation of Figure 6.5 (a). The higher the humidity of air, the faster will be the discharge of the electroscope. Depending upon the value of this humidity, air can behave as a bad or good conductor. It is the presence of water in the humid air which makes it behave as a conductor, due to the fact that water itself is a good conductor for the usual experiments of electrostatics, as discussed in Section 7.11.

Experiment 7.27

There is another easy procedure which can affect the insulating property of air. We repeat Experiments 6.2 and 7.25 in a dry day. The electroscope is initially electrified. Its strip remains lifted, as in Figure 7.42 (a).

(a) (b) (c)

Figure 7.42: (a) Electrified electroscope. (b) The strip drops in just a few seconds by striking a match or a lighter in its neighborhood. (c) The strip remains vertical after removing the match.

We strike a match or a lighter in the vicinity of the electrified electroscope. The strip drops in just a few seconds (Figure 7.42 (b)). The strip does not lift after removing the match (Figure 7.42 (c)).

This experiment shows that the electroscope is discharged very quickly when there is a fire in its neighborhood.

The modern interpretation of this phenomenon is that the flame increases enormously the ionization of air. With the increased number of mobile positive and negative charges in air, the charges over the electroscope are quickly neutralized by these ions. With the increased conductivity of air, the electroscope is also discharged through the hand and the ground. Fire makes the air behave as a good conductor.

7.14 How to Discharge an Electrified Insulator?

The grounding is the easiest way to discharge an electrified conductor, as we saw in Sections 4.5, 6.2, and 6.3. To do this, the conductor needs only to get in touch with the ground. Another procedure is to connect the electrified conductor with the ground through a conducting substance (like the human body or a metal wire).

But we cannot discharge an electrified insulator through this procedure. This is evident in the first experiment which gave rise to the science of electricity, the amber effect, Sections 2.1 and 2.2. In Figure 2.3 we have a rubbed straw attracting small pieces of paper. The plastic straw is kept in our hand while attracting the pieces of paper. Despite this fact, the straw has not been discharged. This is one of the main distinctions between conductors and insulators, as discussed in Section 7.1.

Experiment 7.28

A plastic straw is rubbed in hair. Afterward the straw attracts pieces of paper, as in Experiment 2.1. We then try to ground the straw. To do this, we connect a metal wire between the ground and one of the rubbed portions of the straw. After this procedure, the metal wire is moved away. We move once more the straw near the small pieces of paper. We observe that the straw still attracts these pieces of paper.

What happens in this experiment is simple to describe. Only the specific place of the plastic straw which came into contact with the metal wire was discharged to the ground. That is, the other rubbed portions of the straw are not discharged through this electrical grounding. After all, these charges cannot move through the straw due to the fact that plastic is a very good insulator.

In the next experiment we will see three procedures utilized to discharge an electrified insulator.

Experiment 7.29

Three plastic straws are equally rubbed in hair. Let us call them straws *I*, *II*, and *III*. After being charged, straws *II* and *III* are kept vertically on separate appropriate supports, like those of the electric pendulum. For instance, they can be supported over the paper fasteners connected to a plastic coffee cup filled with gypsum dough or white cement (Figure 7.43). With straw *I* we repeat Experiment 2.1 and observe it attracting small pieces of paper when brought close to them.

After some minutes or a few hours, this experiment is repeated with straw *II*, which received only the initial friction together with straws *I* and *III*. Normally it will attract a very little number of pieces of paper, clearly less than the number attracted by straw *I*. The number of pieces of paper will depend upon the waiting period after the initial friction, the kind of friction, the sort of plastic and the local weather (that is, if it is a dry or a humid day).

In the next day this experiment is repeated with straw *III*. Nothing has been done with this straw after the initial friction. We observe that it no longer attracts small pieces of paper (Figure 7.43).

The symbol F in Figure 7.43 indicates only that the straw has been rubbed several hours ago. Nothing else has been done with this straw. Despite this friction, it looses its electrification after a long waiting period. That is, it looses its ability of attracting small pieces of paper when brought close to them. By the definition of Section 2.1, this means that the straw has become once more electrically neutral, as it was before friction. That is, it lost its electrification, or the charges it had acquired during the friction procedure. These charges are

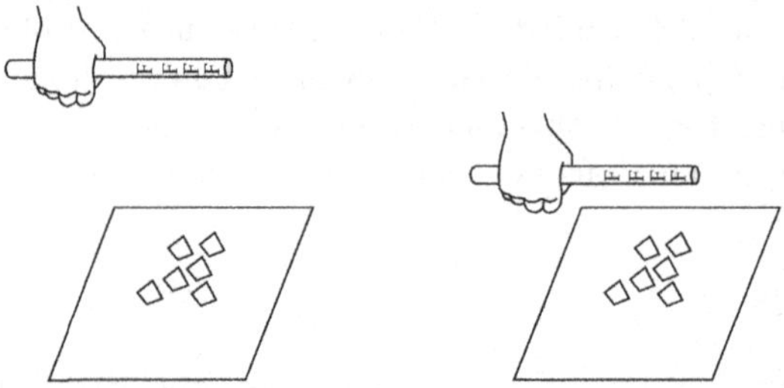
(a) (b)

Figure 7.43: A rubbed straw looses its electrification several hours after the initial rubbing. It no longer attracts small pieces of paper when brought close to them.

lost to the surrounding air. Although dry air is a good insulator, it is not a perfect insulator, as discussed in Subsection 6.7.1.

Experiment 7.30

We now present the second procedure to discharge an electrified insulator.

Initially we rub in hair a plastic straw. It attracts pieces of paper, as in Experiment 2.1. After this trial, the plastic straw is submerged in a receptacle full of fresh water. The straw is removed from the water. We again bring the straw close to the pieces of paper. This time it does not attract them any longer, as in Figure 7.43. As in the other experiments, we should only bring the straw close to the pieces of paper, avoiding them from coming into contact. If they come into contact, the pieces of paper may stick to the straw due to its humidity. We can also give a few knocks upon the straw after it has been removed from the water, or blow it lightly, in order to eliminate the excess of water over its surface.

This experiment indicates that the rubbed straw lost its attracting power after being submerged in water. This means that the water neutralized the straw. This neutralization is again due to the conducting power of fresh water. When water comes into intimate contact with all rubbed portions of the plastic straw which were electrified by rubbing, it neutralizes these surface charges due to the great number of its mobile positive and negative ions. The difference in comparison with Experiment 7.28 is that now we have a grounding of all electrified portions of the rubbed straw, which are then simultaneously neutralized.

Experiment 7.31

The third procedure to discharge an electrified insulator is also very simple.

Initially we charge a plastic straw by rubbing it in hair. It then attracts pieces of paper, as in Experiment 2.1. We remove the rubbed straw far away from the table. After this procedure, we strike a match or a lighter in the

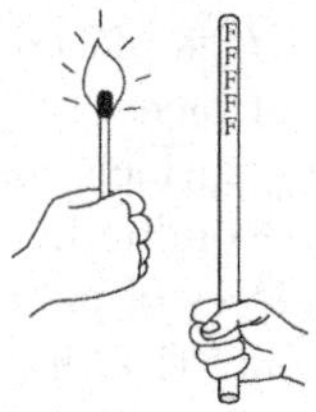

Figure 7.44: We strike a match near a rubbed straw.

neighborhood of the straw. The flame should move near all portions of the rubbed straw, as in Figure 7.44.

After this procedure, the straw is brought again close to the pieces of paper. This time it does not attract them any longer, as in Figure 7.43. The straw has been discharged with this procedure.

As we saw in Section 2.6, Gilbert was the first to describe the phenomenon. He observed that a flame prevented the usual attractions exerted by rubbed substances. Instead of observing the attraction or lack of attraction exerted upon pieces of paper, he analyzed whether or not rubbed amber oriented a nearby versorium. As he put it, "[rubbed electrics] have no effect on a versorium if it have very near it on any side the flame of a lamp." He interpreted this fact supposing that the flame consumed the supposed effluvia emitted by rubbed substances. He believed that the usual attractions exhibited by rubbed amber were due to the action of these emitted effluvia.

Nowadays we have another interpretation for this phenomenon. What happened in this case is analogous to the situation described in Experiment 7.27. That is, the flame increases enormously the ionization of the air molecules. Therefore, air begins to behave as a good conductor, having now a great number of positive and negative mobile ions. The intimate contact between this ionized air and all portions of the rubbed straw neutralizes the charges which were located over the whole surface of the plastic. In this way the straw looses the charges it had acquired by friction. Therefore, it no longer attracts the small pieces of paper.

7.15 A Small Piece of Paper is Attract with a Greater Force when Above an Insulator or a Conductor?

Experiment 7.32

We choose a plastic bag which behaves as an insulator, i.e., which does not discharge an electrified electroscope when the two are touched. We cut several pieces of this plastic bag and several pieces of a sheet of paper. We then create two surfaces: an insulating surface consisting of a Styrofoam plate and a conducting surface consisting of a sheet of paper (or a table or metal plate).

A group of pieces of plastic, group I, is placed on the Styrofoam plate. On the same plate we also place a group of pieces of paper, group II, separated from the first group. On the conducting surface, another group of pieces of plastic, group III, is placed. On the same conducting surface, but separated from the third group, we place a group of pieces of paper, group IV.

Before beginning the experiment, it is important to verify that the small pieces of plastic on the Styrofoam plate and sheet of paper are really neutral. Sometimes these pieces of plastic become electrified during their manipulation, while we cut them, etc. To verify this charge neutrality, we bring a neutral plastic straw close to these small pieces of plastic. When there is no attraction between them, the pieces of plastic can be considered neutral. When they are attracted by the neutral straw, this means that the pieces of plastic are electrified. If this happens, we need to wait for several hours until they discharge naturally through the air. We can then begin the experiment.

We now rub a plastic straw in hair. We bring the horizontal straw close to each of these four groups. We observe that the greatest force is exerted upon group IV, the pieces of paper on the conducting surface, followed by group II, the pieces of paper on the insulating surface. Groups I and III, the pieces of plastic on either surface, are very weakly attracted, and it is difficult to distinguish which one undergoes the greatest attraction. The strength of the force can be estimated in two ways. The first is by observing the distance at which the force begins to be detected, as indicated by the motion of the pieces of paper or plastic. The greater the distance, the stronger the force (Figure 7.45). The second procedure is to observe the number of pieces of paper or plastic that are attracted by the rubbed straw when it is at a constant distance from the table (such as 5 or 10 cm).

Experiment 7.33

We now use an electric pendulum consisting of a paper disk supported by a silk thread. We rub a plastic straw I in hair, so that it becomes negatively charged. It is then supported on an appropriate support far away from the pendulum. A second straw is charged positively by rubbing between two hard rubber hoses. It is also supported far away from the pendulum and from the first straw.

We touch the paper disk with our finger. We then slowly bring straw I near it. The pendulum is attracted by the straw. We do not allow them to come into contact. After removing the first straw, we bring the second straw near the pendulum. The pendulum is also attracted by this straw. We also prevent them from coming into contact.

We now hold a sheet of paper (or metal plate) and place it at an angle below the paper disk on the side opposite from the straw (Figure 7.46). The sheet of paper should touch the disk.

We then bring the first straw near the pendulum in such a way that the pendulum remains between the paper sheet and the straw. At a certain distance the disk moves toward the straw, moving away from the sheet of paper. When

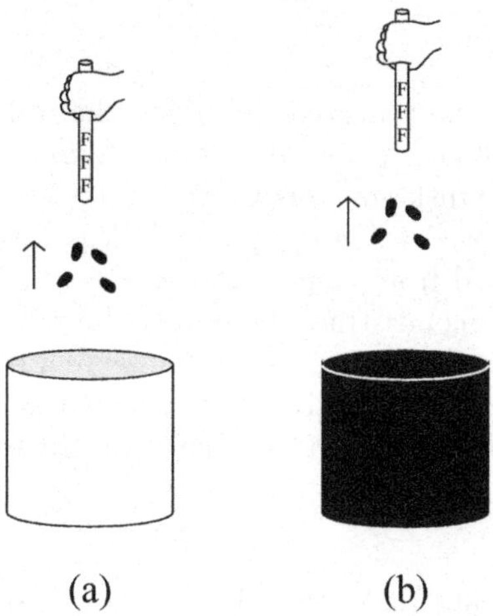

Figure 7.45: (a) Pieces of paper supported by an insulating surface being attracted by a rubbed straw. (b) Pieces of paper supported by a conducting surface being attracted by a rubbed straw. The minimum distance at which the straw can attract the pieces of paper is larger in case (b) than in case (a). The arrows indicate the direction of motion.

this happens, we remove the sheet of paper, but preventing the disk and straw from coming into contact. After this the straw can also be removed and the pendulum will return to the vertical. If we now again move straw I toward the pendulum, we will see an attraction between them. They must be prevented from coming into contact.

We then remove the first straw and bring straw II slowly toward the pendulum. It should not be brought too close to the disk and we must watch attentively in which direction the disk tends to move. When this experiment is performed carefully, we observe that the disk tends to move away from straw II, as it is repelled by it! This indicates that both bodies have charges of the same sign: positive. If straw II moves too close to the disk, there will be an attraction between them, as seen in Section 7.10. This should be prevented.

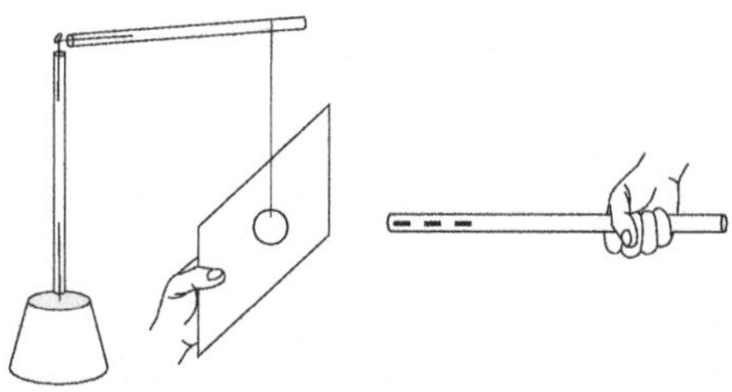

Figure 7.46: Initially the paper disk of an electric pendulum touches the sheet of paper or metal plate.

Experiment 7.34

Experiment 7.33 can be repeated, inverting the order in which the straws are brought together. That is, initially straw *II*, which is positively charged, is moved near the paper disk which is touching the sheet of paper. This sheet of paper should be removed when the paper disk is moving toward the straw. They should be prevented from coming into contact. After the sheet of paper has been removed, the second straw should also be removed. If we now slowly bring straw *I* toward the paper disk, we will observe a repulsion between them. This indicates that both have negative charges. That is, the charge acquired by the paper disk is of opposite sign to the charge on the second straw.

Experiment 7.35

We repeat Experiments 7.34 and 7.35, but now supporting the paper disk upon a Styrofoam plate. At the end of this experiment we observe that the paper disk is attracted by both straws, *I* and *II*, no matter which one of them was first moved toward it.

Experiments 7.33 and 7.34 show that when a piece of paper (in this case the paper disk), supported upon a conducting sheet of paper, is attracted by an electrified straw, there is an electrification of the piece of paper. Moreover, the charge it acquires after touching and being moved away from the sheet of paper is of opposite sign to the electrified straw. Experiment 7.35, on the other hand, shows that a piece of paper, when supported by an insulating Styrofoam plate, does not acquire any net charge after being attracted by an electrified straw. That is, there is no net charge upon the small piece of paper after it is moved away from the Styrofoam plate.

Chapter 8

Final Considerations

8.1 Changing Names and Meanings: From Electric and Non-Electric Bodies to Insulators and Conductors

In 1600 Gilbert had classified the bodies as electric and non-electric, Sections 2.7 and 2.8. The electric materials like amber were those which, after being rubbed, acquired the property of attracting light substances near them. The non-electric materials did not acquire this property with friction. All metals, in particular, were among non-electric bodies.

In 1729 Gray discovered that, by placing an electrified body like rubbed flint-glass in contact or close to a metal, the metal acquired the property of attracting nearby light substances. We saw experiments showing this effect in Section 7.4. Gray's procedures will be discussed in great detail in Appendix B. The same attractive behaviour happened also with other bodies which were classified as non-electric in Gray's time. Du Fay discovered the *ACR* mechanism, showing that a thin piece of metal did become electrified by coming into contact with another rubbed body, Section 4.8. Later on, people did learn how to electrify a piece of metal by induction utilizing the electric polarization and the electrical grounding, Section 7.5. In the 1770's it was discovered that metals could also be electrified by friction, provided they were insulated from the Earth, Section 6.8. These discoveries led to the abandonment of Gilbert's nomenclature and classification scheme.

Since then we have adopted another classification. Substances are now classified as *conductors* and *insulators*. These expressions are due to Du Fay and Desaguliers, Subsection 6.3.1. Most bodies which Gilbert classified as electrics are now called insulators. The bodies which were called non-electric are now called conductors. There is a conceptual novelty here. It is not only a change of names or a simple change of nomenclature. After all, it is possible to make metals attract light substances after being rubbed, provided the metals are insulated during friction. The same happens with other insulated materials. It is Gilbert's distinction which makes no sense any longer. What characterizes the ideal conductors is the fact that they have mobile electric charges and allow the passage or flow of charges through them. Ideal insulators, on the other hand, have no mobile charges, except inside their molecules. Moreover, insulators do not allow the passage or flow of charges through them.

There is a gradation between good conductors and good insulators. Moreover, these properties depend not only upon intrinsic properties of these bodies, but also upon the external conditions to which they are subjected. In any event, the distinction between conducting and insulating substances is one of the most important characteristics in the whole science of electricity.

8.2 Simple and Primitive Facts about Electricity

After performing the experiments described in this book we have reached a reasonable knowledge about electricity. Obviously we did not cover all possible aspects of the subject. In any event, we now have a precise notion of the main facts of electricity. In this Section we present—to use Du Fay's words—the simple and primitive facts or principles about electricity.[1] Here we only describe these observed facts, they are not explained. As a result we can consider them to be primitive. That is, we can utilize these simple principles in order to explain other phenomena and also to explain more complicated experiments, but the fundamental principles themselves are not explained. It is never possible to explain everything. It is always necessary to start with—or to assume as true—some initial facts. We then utilize these primitive assumptions to explain other observations of nature. Here are the primitive facts:

1. The bodies of nature can be found in three different states called electrically neutral, positively charged, and negatively charged. It is also said that they have null charge, positive charge, and negative charge, respectively. We can also say that the bodies are not electrified, are electrified positively, and negatively.

2. These states are characterized by the observed behaviour of bodies. Two neutral bodies neither attract nor repel one another. There is an attraction between a positive body and an initially neutral body. There is also an attraction between a negative body and an initially neutral body. Bodies having charges of opposite sign attract one another. Bodies with charges of the same sign normally repel one another, but in some situations they can also attract one another.

3. These forces of attraction and repulsion increase in intensity when the distance between the interacting bodies decreases. The intensity of these

[1] [DF34b, p. 525].

forces also increases when the strength of charge in the bodies increases. These forces are mutual, acting with the same intensity on both interacting bodies. They are directed along the straight line connecting the bodies, although in opposite directions.

4. The bodies can be divided into two groups called conductors and insulators. The main difference between these two groups is that conductors have mobile charges and allow the passage or flow of electric charges through them. Insulators, on the other hand, have no mobile charges, except inside their molecules. Insulators do not allow the passage or flow of charges through them.

5. The conductors and the insulators can be electrically neutral, positive, or negative. When a charged conductor touches the ground, it discharges. This process is called grounding. The same discharge does not happen for a charged insulator touching the ground. Another way of performing this classification is to touch one end of the body in the cardboard of an electrified electroscope and to touch another end of the body with the ground. The bodies which discharge the electroscope are called conductors, while the bodies with do not discharge the electroscope are called insulators.

6. A body which behaves as an insulator when under a small electric potential difference may behave as a conductor when this potential difference increases beyond a certain value. The majority of solid and liquid bodies behave as conductors in the usual experiments of electrostatics, as few of them are insulators. Among insulators we can mention dry air, amber, silk, and most plastics and resins.

7. The behaviour of a body as a conductor or as an insulator depends also upon other aspects. Let us suppose that one end of the body touches the cardboard of an electrified electroscope, while another end of the body touches the ground. The factors which influence upon the properties of this body are the following: (a) The time required to discharge an electroscope (the greater the time of contact, the greater will be the amount of discharge). (b) The length of the body (the greater this length, the slower will be the discharge). And (c), the cross-sectional area of the body (the greater this area, the faster will be the discharge).

8. Neutral bodies can be charged by several mechanisms. The most common procedure is friction of two neutral bodies. After the friction, one of the rubbed bodies becomes positive and the other negative. The insulators are only charged on the rubbed portion of their surfaces. The charge acquired by the rubbed conductors, on the other hand, spreads over the outside surfaces of conductors when conductors are completely surrounded by insulators, or go to the ground if there is a conducting contact with the Earth.

9. A neutral conductor can also acquire a charge from a charged insulator when they are put into contact with one another, without any friction. The charge acquired by the conductor has the same sign as the charged insulator. In this process the amount of charge lost by the insulator is equal to that gained by the conductor. On the other hand, the amount of charge acquired by a neutral insulator when it touches another charged insulator is negligible when there is no friction between them.

10. Conductors polarize electrically in the presence of a nearby charged body. The portion of the conductor which is closest to the charged body becomes electrified with a charge having a sign opposite that of the nearby charged body. The farthest portion of the conductor becomes electrified with a charge of the same sign as the nearby body when the conductor is electrically insulated. If the conductor is insulated and if it be broken apart in these two parts in the presence of the nearby charged body, the two parts will become electrified with charges of opposite sign.

11. If the conductor is electrically grounded in the presence of the nearby charged body, the portion of the conductor which is farthest from the charged body will be neutralized. This fact allows a conductor to be electrified with a charge of opposite sign to the nearby body.

12. The molecules of an insulator are polarized in the presence of a nearby charged body. The portion of any molecule which is closer to (farther from) the charged body becomes electrified with the opposite (same) sign as the charged body. These polarized charges are restricted to the molecules and do not move along the insulator. Moreover, they do not pass to another conductor which comes into contact with the insulator.

13. The number of polarized charges in conductors close to a charged body increases when the distance between them decreases. The same happens with the effective polarized charges of insulators close to a charged body.

14. There is a higher polarization of conductors and insulators when the degree of electrification of the nearby charged body increases.

15. A force of non-electrostatic origin keeps the charges on the surfaces of conductors and insulators at rest when these bodies are electrified or polarized. A force of non-electrostatic origin is also responsible for generating opposite charges when two bodies are rubbed against one another.

When describing these simple facts, we should bear in mind that we are talking in general terms, referring implicitly to the experiments described in this book. All these effects depend on the order of magnitude involved in the experiments, there are always exceptions in all experimental descriptions. For instance, when we say that two neutral bodies do not interact with one another, we are not considering the gravitational attraction between them. The reason is that this gravitational interaction is not observed or cannot be detected in ordinary experiments involving small, light bodies. It shows its effect only when at least one of the bodies is of astronomical dimensions, like our own Earth. When we say that a charged body attracts a body which is initially neutral, it is assumed we are dealing with light bodies or bodies supported by strings in such a way that there is only small resistance to the lateral motion of these bodies. If this is the case, these neutral bodies will be able to move near the charged body when there is an attraction between them. Moreover, in order to observe this effect of attraction, the distance between the interacting bodies cannot be very large and the charge on the rubbed body should not be very small, otherwise these effects are not perceptible. The same applies to the other principles.

8.3 Description of the Amber Effect

These primitive principles can be utilized to explain or describe more complex phenomena. But here we will use them to describe what happened in Experiment 2.1, which is analogous to the amber effect, the first experiment in the history of electricity. In this experiment an unrubbed plastic straw did not attract small pieces of paper, while a rubbed straw attracted these pieces of paper when moved near them, as in Figures 2.1 and 2.3. Experiment 2.11 yielded a different result. In this case a wood skewer, rubbed or unrubbed, did not attract pieces of paper (Figure 2.9). In Experiments 2.3, 7.18, and 7.20 we observed that a plastic rubbed straw exerts a greater force upon small pieces of paper than upon small pieces of plastic of approximately the same weight and shape as the paper. In Experiments 7.32 to 7.35 it was shown that the pieces of paper resting upon a conducting surface experience a greater attraction from a nearby rubbed straw than the pieces of paper resting upon an insulating surface. Moreover, the pieces of paper which were resting above a conducting surface acquire a net charge after being attracted by a nearby charged body. The net charge acquired by the pieces of paper have an opposite sign to the sign of the nearby body which attracted them.

In these experiments the bodies behaving as insulators were air, the rubbed plastic straw, the silk thread of the electric pendulum, the Styrofoam plate and the small pieces of plastic which were being attracted. The plastic straw was rubbed with hair. From what was seen in Section 5.4, the straw acquired a negative charge. Even though it was being held in our hand, it did not discharge because plastic is an insulator.

The bodies behaving as conductors in these experiments were the ground, the human body, the wood skewer, the paper disk of the electric pendulum, the sheet of paper which supported this paper disk above it and the small pieces of paper being attracted by the straw. It was not possible to attract the pieces of paper with a rubbed skewer. When the skewer was rubbed, it may have momentarily acquired an electric charge. But since it behaves as a conductor in this situation and was held in our hand in contact with the Earth, the wood was grounded. As a result, any charge appearing on the skewer during the rubbing process would immediately be neutralized. Due to this fact, it did not attract the small pieces of paper, even after being rubbed. A conductor can only keep the charges it acquires by friction when it is insulated, as seen in Section 6.8.

The plastic straw could be rubbed with insulators (hair, a silk cloth, or a plastic bag) or with conductors (our hand, a sheet of paper, or a cotton cloth). It is not crucial to know if this substance was an insulator or a conductor. But the sign of the charge acquired by the plastic straw will depend upon the type of material with which it was rubbed, i.e., an insulator or a conductor.

The attraction the rubbed straw exerted upon a piece of plastic can be illustrated microscopically utilizing a plastic pendulum. As aways, care should be taken in order to utilize a plastic disk cut from a plastic bag which behaves as an insulator. Moreover, this plastic disk should be neutral when it is far away from other electrified bodies. In this case, when we move the rubbed straw near the plastic disk of the pendulum, we observe a small attraction between them. It is assumed that the plastic molecules are polarized in the presence of the rubbed plastic. The portion of each molecule which is closer to the rubbed body becomes electrified with a charge of opposite sign. The portion of each molecule which is farthest from the rubbed body becomes electrified with a charge of the same sign as this body. This is illustrated in Figure 8.1 (a). The interior of the polarized plastic behaves macroscopically as if it were neutral, due to the cancelation of the nearby charges of opposite signs. But the surface of the polarized plastic behaves as if it had an effective charge, as illustrated in Figure 8.1 (b).

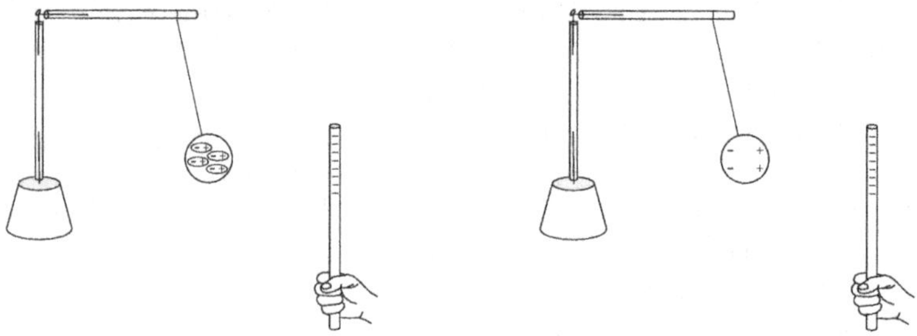

(a) (b)

Figure 8.1: (a) A rubbed plastic polarizing and attracting a plastic disk. (b) Effective polarization of the plastic disk.

It is assumed that the plastic molecules are polarized by to the presence of the nearby rubbed straw due to the existence of positive and negative charges in each molecule. Moreover, these positive and negative charges must be mobile inside each molecule. The polarization of each molecule is also due to the fact that charges of the same sign repel one another, while charges of opposite sign attract one another. Some force of non-electrostatic origin prevents these polarized charges from moving indefinitely far away from one another. The stronger the charge on the electrified plastic straw, the stronger will be the polarization of the plastic disk. This polarization also increases when the distance between the straw and the disk decreases. The greater this effective polarization of the plastic disk, the stronger will be the net force upon it.

The polarization of the plastic is not greatly changed if it is supported on a Styrofoam insulating plate or a conducting sheet of paper. The plastic does not receive a net charge in these two cases.

The observed attraction between the rubbed straw and the polarized plastic may be a consequence of yet another property of the electric forces: The attractive and repulsive forces increase in strength when the distances between the charges decrease. Therefore, the attractive force between the rubbed straw and the effective charges spread over the surface of the plastic which is closer to the straw is greater than the repulsive force between the straw and the effective charges of the same sign spread over the surface of the plastic which is farthest from the straw. The sum of these two forces does not go to zero. The attractive force is greater than the repulsive force. These partial forces of differing magnitude generate a net attractive force between the rubbed straw and the polarized plastic.

Let us now analyze the attraction exerted by the rubbed straw upon a conductor such as a small piece of paper. We assume initially that this piece of paper is the paper disk of an electric pendulum suspended by a silk or nylon thread (insulating materials). This case is different from the plastic disk in two ways. The first is that the polarization of a conductor is due to the motion of free charges over the volume of the conductor. That is, there is a real macroscopic polarization of the conductor, not just a polarization of its molecules (Figure 8.2).

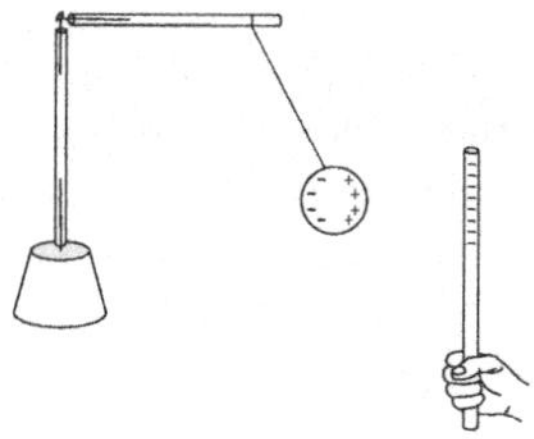

Figure 8.2: A rubbed straw polarizing and attracting a paper disk.

The second difference is that this polarization is of greater intensity than the effective polarization of an insulator in the presence of the rubbed straw. This results in a stronger force exerted by the rubbed straw upon the paper disk than the force exerted by the rubbed straw upon a plastic disk. Here we are assuming paper and plastic disks of the same shape and size, with both of them at the same distance from the rubbed straw. We can notice this second difference by comparing Figures 8.2 and 8.1. The silk thread of the pendulum with a paper disk is more inclined from the vertical than the silk thread of the pendulum with a plastic disk. In both situations there is the same distance between the rubbed straw and the pendulum. This indicates that the force upon the conducting paper is greater than the force on the insulating plastic. Once more, there is a force of non-electrostatic origin which prevents these polarized charges from separating even more from one another, flying off into the air beyond the paper disk. It is this force of non-electrostatic origin which keeps the polarized charges at the edge of the paper disk, preventing them from discharging through the air.

If the paper disk is grounded while polarized, the charges on the disk which are far away from the rubbed straw are neutralized by the Earth's charges due to the grounding. This is analogous to what we saw in Figure 7.30. In this case, this paper disk has a net charge which is different from zero, opposite in sign to the nearby rubbed straw. Therefore, this charged paper disk experiences a stronger force of attraction from the rubbed straw than the force exerted by the rubbed straw upon the polarized disk. The reason is that in the present situation the repulsion which existed before between the rubbed straw and the charges of the same sign which were spread over the farthest surface of the paper disk no longer exists (Figure 8.3).

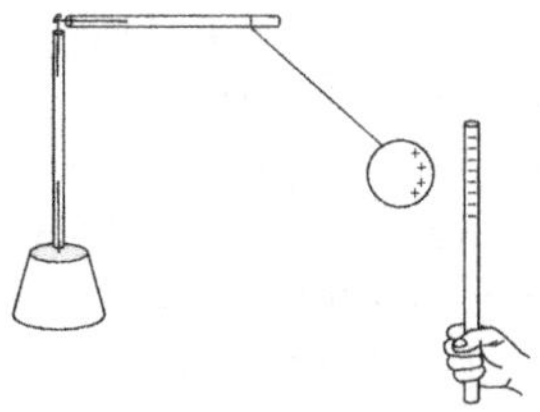

Figure 8.3: An electrified paper disk being attracted by a rubbed straw.

In Figure 8.4 we compare these three cases. We assume that in all cases the rubbed straw is at the same distance from the electric pendulum. In (a) we have a small attractive force exerted upon a plastic disk which is polarized by a nearby electrified straw. In (b) there is a stronger force exerted upon a conducting disk which is more strongly polarized. In (c) there is an even greater attractive force exerted by an electrified straw upon a charged conducting disk.

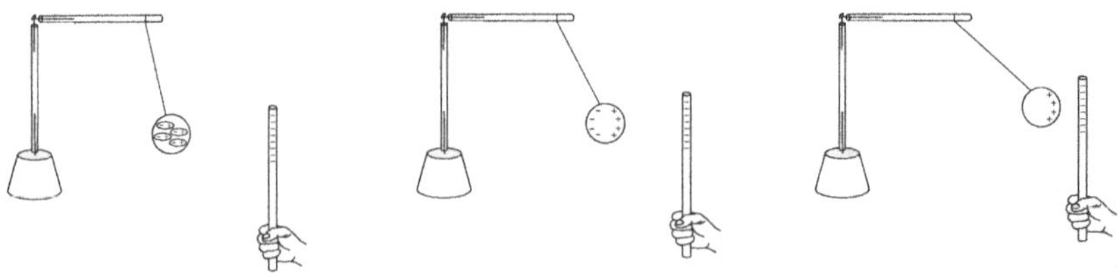

(a) (b) (c)

Figure 8.4: (a) A polarized insulator being attracted by a nearby electrified body. (b) A polarized conductor being attracted. (c) A charged conductor being attracted. The force increases from (a) to (c), with the rubbed straw at the same distance from the electric pendulums.

The microscopic description of what happens with the small piece of paper (a conductor) supported on a Styrofoam plate (an insulator) when we bring an electrified straw near it in Experiment 7.32 is illustrated in Figure 8.5.

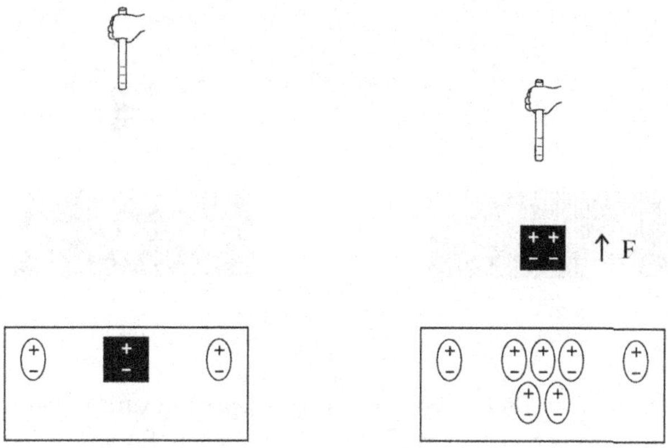

(a) (b)

Figure 8.5: (a) A polarized piece of paper supported on a polarized Styrofoam plate in the presence of an electrified straw which is far away from the paper. (b) By moving the straw even closer toward the paper, we increase the polarization of the paper and the Styrofoam. This creates an attractive electrostatic force upon the paper, which can be larger than its weight. If this happens, the paper moves toward the straw.

The conducting paper is represented by the black rectangle. The insulating Styrofoam is represented by the large white rectangle. When an electrified straw is placed above the paper, but far away from it, the paper and the Styrofoam become polarized (Figure 8.5 (a)). There appears an electrostatic attractive force between the straw and the piece of paper. As this electrostatic force is smaller than the weight of the paper, it does not move toward the straw. By moving the straw even closer to the paper, we increase the polarization of the paper and the Styrofoam. This increases the attractive force between the straw and the paper. When the distance between the straw and the paper is less than or equal to a certain value d_1, the electrostatic force becomes larger than the weight of the paper. The paper then moves toward the straw (Figure 8.5 (b)). The strength of the force is indicated by the size of the arrow.

Different processes take place when the conducting piece of paper is initially supported on a conducting sheet of paper (Figure 8.6).

The conducting piece of paper is represented by the small black rectangle, while the conducting sheet of paper is represented by the large black rectangle. By placing an electrified straw above the paper, but far away from it, we create a redistribution of charges upon the ground. Consequently, the surface of the piece of paper and the surface of the sheet of paper become electrified with charges opposite in sign to the charge on the straw (Figure 8.6 (a)). An electrostatic attractive force is then generated between the piece of paper and the straw.

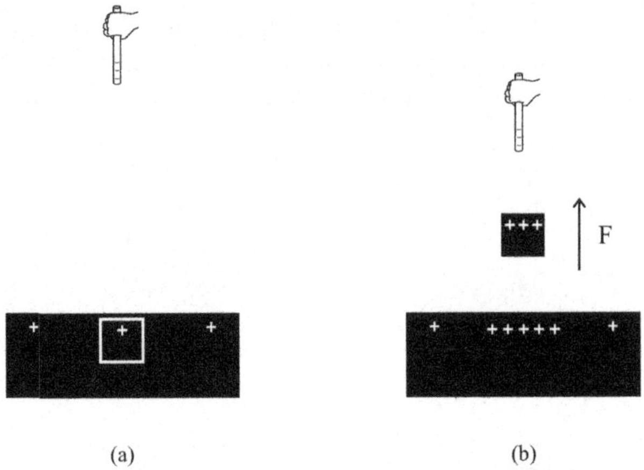

Figure 8.6: (a) An electrified piece of paper supported on an electrified sheet of paper in the presence of an electrified straw which is far away from the paper. (b) By moving the straw even closer to the piece of paper, we increase the electrification of the piece of paper and the sheet of paper. This generates an electrostatic force on the piece of paper which may be larger than its weight. The piece of paper can then move toward the straw.

Because this electrostatic force is smaller than the weight of the piece of paper, it does not move toward the straw. By moving the straw even closer the piece of paper, we increase the electrifications of the piece of paper and of the sheet of paper. When the distance between the straw and the piece of paper is less than or equal to a certain value d_2, the electrostatic force becomes larger than the weight of the piece of paper. It then begins to move toward the straw (Figure 8.6 (b)). This distance d_2 is larger than the previous distance d_1 (Figure 7.45).

Let us suppose that the pieces of paper are moving in air, being attracted by the rubbed straw. We compare here two cases. In the first case the piece of paper was initially supported on an insulating surface. In the second case the piece of paper was initially supported on a conducting surface. We will assume that the straw is at the same distance from both surfaces. We will also assume that both pieces of paper are half-way between the lower surfaces and the straw. The force strengths are indicated by the sizes of the arrows. It is smaller in the first situation than in the second. This can be visualized comparing Figures 8.5 and 8.6 (Figure 8.7). The reason for this difference is that in the second case there is a net charge on the piece of paper, having an opposite sign than the charge on the straw. This is not the case in the first situation. Moreover, there is also a repulsive force upon the piece of paper exerted by the lower conducting surface in the second situation. This is due to the fact that these two bodies have charges of the same sign.

In Experiment 2.1 it is not easy to detect the net charge upon the piece of paper when it is moving toward the rubbed straw, after it is removed from the

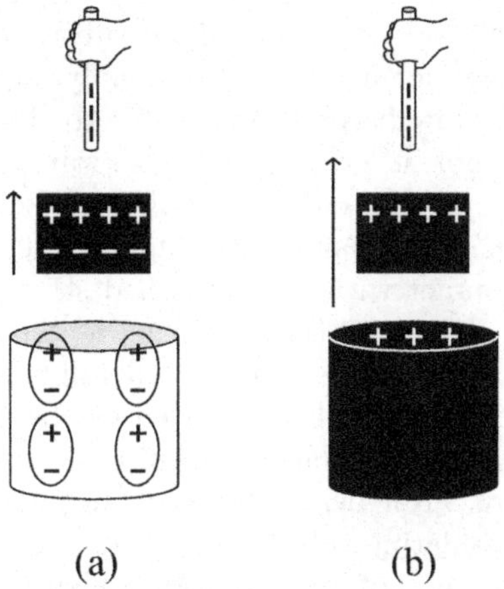

Figure 8.7: (a) Polarized piece of paper, which was initially supported upon an insulating surface, being attracted by a rubbed straw. (b) Electrified piece of paper, which was initially supported upon a conducting surface, being attracted by a rubbed straw. The sizes of the arrows indicate the force strengths, assuming the same distances in both cases.

surface of the ground or a table. However, this can be shown by careful observations, as indicated in Experiments 7.33 and 7.34. In Figure 8.7 we illustrate microscopically what was happening in Figure 7.45.

It is a surprise for many people that in the oldest experiment of electricity, like that of Experiment 2.1, the light substance and the ground behave as conductors. Even more surprising is the fact that, in general, the light substance has a net charge when it is moving toward the rubbed plastic. Despite these surprises, this is the most common situation in which electrostatic attractions are observed. That is, normally when a solid or liquid substance is being noticeably attracted by an electrified body, this substance will be a conductor. If the substance were an insulator, the net force upon it would usually be small, making it difficult to observe its movement. Moreover, as usually this conducting substance which is being attracted was previously supported above another conducting substance (like the ground, the human body, a wooden table, a sheet of paper, or a metal plate), it will have a net charge while it is moving toward the rubbed plastic. And the net charge upon the conducting light substance in this situation will have a sign opposite to the attracting electrified body.

Kelvin's electrostatic generator, Subsection 7.12.1, is analogous to Experiments 7.32 to 7.34. The difference is that it utilized water droplets instead of small pieces of paper.

Here we are not considering the influences of other nearby bodies in the amber effect experiment. Moreover, we are not considering what would happen if there were a conducting or insulating plate between the rubbed straw and the light substances. Nor are we considering the influence of the shape of the support on the net force exerted upon the light substance.

These phenomena may not happen, or may only happen with a very small intensity, if the rubbed straw becomes moist due to the sweat of our hand, or due to humidity of the ambient air. Because fresh water behaves as a conductor in these experiments, it can help to discharge the rubbed straw. This grounding may happen either through our hand or through the surrounding air.

After coming into contact with the electrified straw, the small conducting substance may receive a charge of the same sign as the straw due to the *ACR* mechanism. After this contact, the substance will fall to ground either due to its weight or due to the electrical repulsion exerted by the straw. This mechanism does not work as well when an insulating substance comes into contact with the rubbed straw. Therefore, after being attracted and coming into contact with the rubbed straw, the insulating substance falls to the ground after some time due to its weight and to a loss of electrification which happens naturally in the straw with the passage of time. This loss of electrification is due to the small conductivity of dry air.

As we can see, a great many phenomena and processes take place in the experiment of the amber effect, which is analogous to Experiment 2.1. As a result, we can see why it took so long to achieve a clear perception of what was taking place here. Moreover, we are only describing in more detail the many microscopic processes taking place, but we are not explaining the experiment. After all, we have not explained why charges of opposite sign attract one another, or the mechanisms responsible for electrification by friction, or the reasons why some bodies behave as conductors while others behave as insulators, or the origin of the non-electrostatic forces which keep the charges fixed at the surfaces of electrified or polarized conductors and insulators; nor did we explain the reason why the force depends on distance, the triboelectric series, or the reasons why a certain body A becomes negatively charged by being rubbed against a certain body B (instead of becoming positively charged, or instead of continuing to be electrically neutral), etc.

In any event, nowadays we have a reasonable knowledge of what is happening in many electric phenomena. Moreover, we can control many mechanisms involved in these processes. This certainly represents a great advance in our dominion over nature and in our comprehension of many physical phenomena.

The history of electricity is relatively short compared with astronomy, geometry, or mechanics. This is why we still have access to the writings of some of the main scientists responsible for discovering the most important electrical phenomena. It is fascinating to reproduce their experiments with simple, inexpensive materials. It is also very interesting to read the accounts of their works, and see how they reacted to their findings, what guided them, etc.

With this book we hope to help others to follow the fascinating paths of nature, as discovered by some of the foremost scientists in the early history of electricity.

Appendix A

Definitions

Here we present definitions of some expressions utilized in this book.

Amber - A hard yellowish or brownish translucent fossil resin.

Jet - A compact velvet-black coal that takes a good polish.

Catgut - A tough cord made usually from sheep intestines.

Copal - A recent or fossil resin from various tropical trees.

Flint-glass - A heavy brilliant glass that contains lead oxide.

Lac - A resinous substance secreted by species of scale insect and used chiefly in the form of shellac.

Nylon - Any of numerous strong elastic synthetic polyamide materials that are fashioned into fibers, filaments, bristles, or sheets and used especially in textiles and plastics.

Polyamide - A polymer containing repeated amide groups, a polymeric amide. Nylon is a synthetic polyamide material.

Polyester - Polyester is a category of polymers which contain the ester functional group in their main chain.

Resin - Any of various solid or semisolid amorphous fusible flammable natural organic substances that are usually transparent or translucent and yellowish to brown, are formed especially in plant secretions.

Appendix B

Stephen Gray and the Discovery of Electrical Conduction

One of the most important aspects of the whole science of electricity is the fact that there are two sets of bodies with very distinct properties, *insulators* and *conductors*. In the case of insulators, the charges generated by friction remain at the rubbed region, and do not move along the material. A rubbed insulator is not discharged by coming into contact with the ground. In conductors, on the other hand, the charges generated by friction immediately spread over the whole surface of the conductor. If a charged conductor comes into contact with the ground, it discharges immediately, loosing its electric charge to the Earth.

The discovery of these two types of bodies and their main properties came only very late in the history of electricity. Stephen Gray (1666-1736) made this great discovery in 1729, publishing a fundamental work on the subject in 1731.[1] We present here a few aspects of his life and work.[2] He published some of the most important papers in the early history of electricity.[3]

Gray was born in 1666 in Canterbury, England. There is no known portrait of Gray. His father and his brother were dyers by profession. They dyed cloth during its manufacture. Gray himself also worked as a dyer, as established by Heilbron.[4] He never studied at a university. He was an amateur scientist, making contributions mainly in astronomy and electricity. He probably never married. When he was 53 years old, he began to live as a pensioner at the Charterhouse, a charitable home for retired sea captains and poor boys. The people who lived there led a simple life, with little comfort, although there was no fear of starvation. He lived there until his death at 70 years of age.

[1] [Grah].

[2] [CM79], [Hei81c], [Hei99, pp. 242-249], [CC00], and [BC09].

[3] [Chi54], [Grab], [Grah], [Grad], [Graf], [Grag], [Grae], [Grai], [Grac], and [Graa].

[4] [Hei81c].

He may have studied with his friend, the Astronomer Royal John Flamsteed (1646-1719). In 1707 he was brought to Cambridge by Roger Cotes (1682-1716). Gray was elected a member of the Royal Society in 1732. Due to his researches on electricity, Gray was the first recipient of the Royal Society's Copley medal for scientific achievement.

His interest in electricity was initiated by articles by Francis Hauksbee (born around 1666 and died in 1713) published in the Philosophical Transactions from 1704 to 1707. Hauksbee described experiments with a rubbed glass tube which, in addition to attracting small substances, also emitted light. In 1708 Gray sent a letter to the secretary of the Royal Society, Hans Sloane (1660-1753), describing several experiments on electricity. This letter was only published in 1954.[5] In this letter he described several experiments analogous to those of Guericke in which a down feather is attracted by a rubbed glass, touches it, and is then repelled by it, as in Experiment 4.4. Although Gray did not quote Guericke in his works, he may have known of his experiments.[6] In 1720 Gray published a paper describing new experiments on electricity.[7] In this work he described a kind of electric pendulum and new electric substances. That is, he discovered new substances which attract light bodies when rubbed, or that are attracted by wood or by the human body when these new substances are rubbed, as in Experiment 3.10.

B.1 Gray's Electrical Generator

His main discoveries were made between 1729 and 1736, when he was between 63 and 70 years old. The principal paper describing his discovery of conductors and insulators was published in 1731.[8] Until that time no one had succeeded in making metals attract light bodies, even when they rubbed, heated, or stroked the metals. This was one of the most important discoveries by Gray: he learned to communicate the electrical property of attracting light bodies to a large variety of substances which until then no one had been able to electrify. Gray did not electrify metals by friction. But he showed that metals acquire the property of attracting light bodies when they are connected to a rubbed flint-glass tube, or by simply moving this tube near the metal.

His paper begins with the following words:[9]

> In the year 1729 I communicated to Dr. Desaguliers, and some other gentlemen, a discovery I had then lately made, shewing that the electrick vertue of a glass tube may be conveyed to any other bodies, so as to give them the same property of attracting and repelling light bodies, as the tube does, when excited by rubbing; that this

[5][Chi54].

[6][Chi54, p. 38, Note 6].

[7][Grab].

[8][Grah].

[9][Grah, pp. 18-19].

> attractive vertue might be carried to bodies that were many feet distant from the tube.

The hollow glass tube he utilized was of flint-glass, that is, a heavy brilliant glass that contains lead oxide. He rubbed his tube with his bare hand, as he mentioned in his paper of 1707-1708:[10]

> The glass tube made use of was about the size of that made use of by Mr Hauksbee but insted of rubing it with paper as he directs I found it to succeed better with me when rubed with my bare hand only.

He also held the rubbed glass tube in his hand during the experiments. As the tube was not discharged by contact with the hands, this means that the tube acted like a very good insulator, contrary to what happens with most modern glass found in the home.

It is normal for our hands to sweat due to the heat developed during the rubbing process. The glass can become humid during this process, losing some of its insulating properties. Gray's flint-glass was a hollow cylindrical tube, 1 m long. Its length may have been helpful in maintaining its insulating properties. Perhaps he rubbed it only at one end, holding it at the other end. This would keep a reasonable amount of dry glass between these two regions, resulting in a reasonable degree of insulation.

Gray's glass-tube was not only an excellent insulator. From what we will see, he succeeded in transmitting attractive power to very long conducting cords. The glass touched the cord or was kept close to one end, with the other end of the cord attracting brass leaf. This means that he was able to create a strong polarization of the cord due to the large amount of charge in his glass tube. His glass tube was able to accumulate a large quantity of electrical charges during the rubbing process.

He described the tube as follows:[11]

> Before I proceed to the experiments, it may be necessary to give a description of the tube: Its length is three feet five inches [1 m], and near one inch two tenths in [external] diameter [3 cm]: I give the mean dimensions, the tube being larger at each end than in the middle, the bore about one inch [2.54 cm]. To each end I fitted a cork to keep the dust out when the tube was not in use.

This last precaution may have been motivated by experiments Hauksbee had done which showed that contaminants inside the tube might reduce its electricity.[12]

This hollow glass tube rubbed with his hands was his standard electrical generator.

---

[10] [Chi54, pp. 34 and 37].
[11] [Grah, p. 20].
[12] [Haub] and [Hei99, p. 245].

B.2 The Discovery of Electrification by Communication

We now come to Gray's great discovery, made in February 1729 (our emphasis in italics):[13]

> The first experiment I made, was to see if I could find any difference in its attraction, when the [rubbed] tube was stopped at both ends by the corks, or when left open, but could perceive no sensible difference; but upon holding a down-feather over against the upper end of the tube, I found that it would go to the cork, being attracted and repelled by it, as by the tube when it had been excited by rubbing. I then held the feather over against the flat end of the cork, which attracted and repelled many times together; *at which I was much surprized, and concluded that there was certainly an attractive vertue communicated to the cork by the excited tube.*

That is, he had rubbed only the glass tube, but not the cork. On the other hand, he observed that the feather was attracted and repelled by the cork which was in contact with the tube. He made a test and concluded that this was really happening, as the cork attracted not only by its lateral surface, which was in contact with the tube, but also by its plane face which had not been rubbed and which was not in direct contact with the glass.

As Gray did not provide any drawings in his papers, it is not easy to know exactly the kind of experiment he performed here. We see three possibilities.

(a) He may have held the feather quill in his hand, with the quill working as an insulator. He would then have observed the feather bending and being attracted and repelled by the cork, alternately touching the cork and his hand. The fibres of the feather would work as a conductor, being charged by the *ACR* mechanism and then afterwards discharged when they touched his hand, with this process being repeated a few times (Figure B.1).

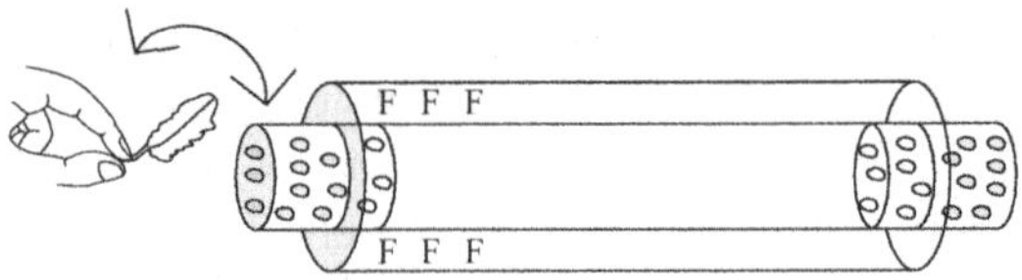

Figure B.1: First possible way Gray may have made his crucial observation.

(b) The feather may have been tied to a silk thread, an insulator, moving like an oscillating pendulum. That is, it would be charged when in contact with the cork and discharged when in contact with a nearby conductor, such as his hand, a wooden object, or a wall (Figure B.2). The verb he used in his description was *holding*. This suggests that he held the feather with his hand. As the feather

[13] [Grah, p. 20].

was attracted and repelled many times altogether, this suggests that the feather was between the cork and a conductor (maybe Gray's hand, a wall, or another object). When the cork was electrified or polarized by the rubbed glass tube, it attracted the feather. After the feather touched the cork, it acquired some charge and was repelled by the cork due to the *ACR* mechanism. The feather could then be discharged in another nearby conductor. After this discharge, it would once again be attracted by the electrified or polarized cork, and this process might be repeated a few times. That is, something analogous to what we saw in Experiment 4.15. As we saw in Section 4.6, in 1720 Gray himself had used an electric pendulum with a feather tied to a silk thread.[14]

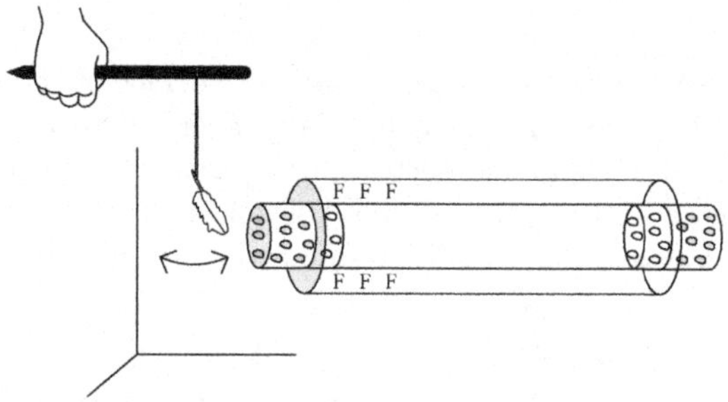

Figure B.2: Second possible way Gray may have performed his crucial observation. In this case the feather would be attached to a silk thread, an insulator.

(c) The third possibility is that the present experiment was analogous to the experiment Gray had performed in 1708 and which we described in Section 4.2. That is, the feather might had been released in air above the cork. It would then be attracted by the cork at the end of the rubbed glass, would have been electrified by the *ACR* mechanism, and then repelled by the cork. If the feather moved near another nearby conductor (like Gray's hand, a wall, or another conductor), it would be attracted by this conductor. It would be discharged upon contact with this conductor, and again be attracted by the electrified or polarized cork. This process might be repeated many times (Figure B.3).

This third possibility seems the most probable to us. The verb *held* had already been utilized by Gray in his second experiment of 1708 described in Section 4.1, p. 67. In that case, after the down feather had been let go from the fingers and was attracted by the rubbed glass tube, if it were held at a short distance from an object, it would oscillate between this object and the glass. We believe that this experiment of 1729 was analogous to the experiment described in Figure 4.11. The difference is that now the feather would oscillate between the cork and a nearby body, with the cork connected to the rubbed glass tube, although the cork itself was not rubbed.

It is also not clear if the rubbed glass tube was vertical or horizontal. Even for a horizontal tube, we can talk of its "upper end" as meaning the portion

[14][Grab].

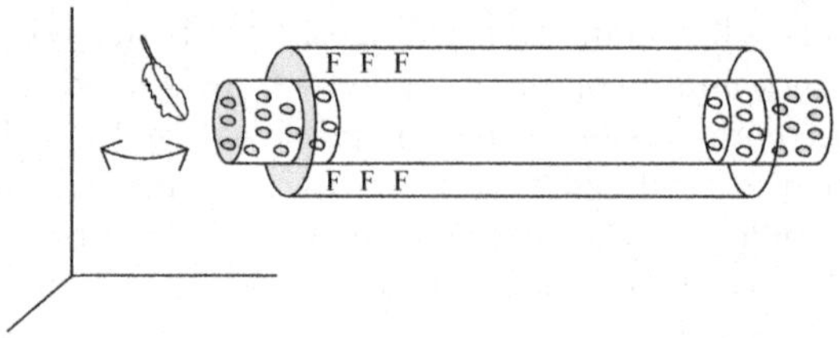

Figure B.3: Third possible way Gray might have performed his crucial observation. The feather would oscillate in air between the cork and another nearby body.

near the rubbed end which was farthest from the surface of the Earth, while the lower end would be the portion near the rubbed end which was closest to the Earth's surface. It may also have been that in part of the experiment the tube was vertical, while in another part it was horizontal.

Although this was a casual discovery (in Gray's words) "*at which I was much surprized*," Gray actually expected that the electricity might be transmitted to other bodies. He had performed previous experiments in which he observed light being emitted from rubbed bodies and going toward other bodies which had not been rubbed, just when these unrubbed bodies were brought near the rubbed ones. Just before describing the previous experiment of the feather and cork, Gray said the following in his 1731 paper:[15]

> I then resolved to procure me a large flint-glass tube, to see if I could make any farther discovery with it, having called to mind a suspicion which some years ago I had, that as the tube communicated a light to bodies, when it was rubbed in the dark, whether it might not at the same time communicate an electricity to them, though I never till now tried the experiment, not imagining the tube could have so great and wonderful an influence, as to cause them to attract with so much force, or that the attraction would be carried to such prodigious distances, as will be found in the sequel of this discourse.

The importance of this discovery is that the cork behaves as a conductor, as we saw in Subsections 6.3.1 and 6.3.2. For this reason it is not possible to charge it by friction while holding it with our hand. That is, any charge it may have acquired by friction would immediately be discharged through our body. For this reason, until then no one had succeeded in causing corks, metals, etc. to attract light bodies after being rubbed, as was easily the case with amber or with a flint-glass. It was the detail of the feather being attracted by the cork which caught Gray's attention. This observation indicated to him that he could somehow transmit the electrical virtue to the cork, which was classified as a non-electric material. This was the first fundamental discovery by Gray in

[15] [Grah, pp. 19-20].

this article: to communicate electricity to another body (like the cork) without rubbing it.

The modern interpretation or microscopic description of the "attractive virtue" which Gray succeeded in transmitting to the cork, is that it became polarized, as in Experiment 7.9. This is illustrated in Figure B.4.

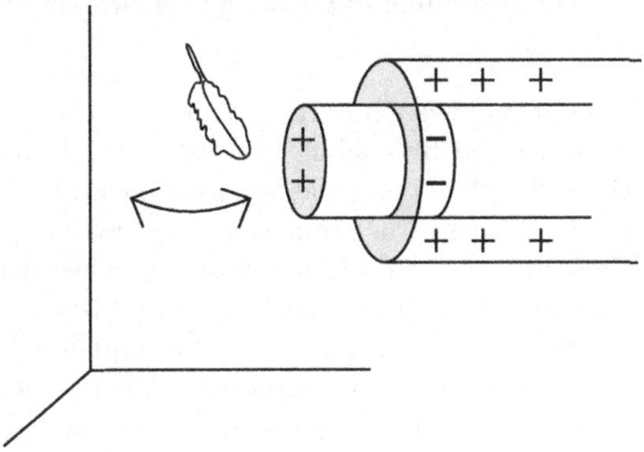

Figure B.4: Polarization of the conducting cork due to the rubbed glass. The conducting feather was attracted by the charges spread over the external surface of the cork.

That is, the insulating glass tube was charged by friction. The conducting cork attached to the rubbed glass became polarized. Its internal surface acquired a charge of opposite sign to the rubbed tube. Its external surface acquired a charge of the same sign as that of the rubbed tube. The nearby conducting feather was attracted by these charges spread over the external surface of the cork.

B.3 Exploring the Discovery and Awakening the Hidden Electricity of Metals

After this casual discovery, Gray continued his experiments.[16] He began to determine systematically to which bodies he could communicate "electricity" or the "attractive virtue." He also wanted to know how far away could he carry these properties. He attached an ivory ball, 3.3 cm diameter, with a hole through it to a wooden stick 10 cm long. The other end of this stick was set in the cork connected to the glass tube. When he rubbed the tube, he observed that the ball attracted and repelled the feather more vigorously than the cork had (Figure B.5). He increased the length of the stick to 20 cm and later on to 60 cm, and the attraction remained. He replaced the wood stick by iron and brass wires, observing the same effects.

[16] [Grah].

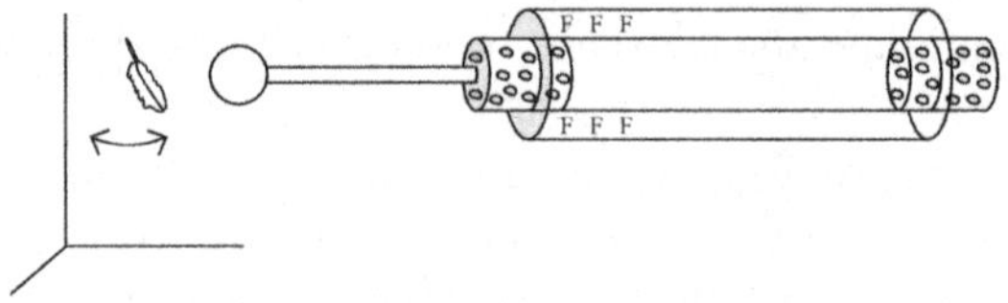

Figure B.5: The beginning of Gray's systematic experiments.

He increased the lengths of the wires up to 90 cm, but then encountered many vibrations. These were caused by rubbing the tube, which made it difficult to observe the attractions. He then hung the ball by a packthread suspended from a loop on the tube. Packthread was stout cord used for wrapping packages.[17] When he rubbed the tube, the ball attracted and repelled a brass leaf placed under it. The same happened for a cork ball and for a 570 g iron ball connected to the packthread (Figure B.6 (a)). Figure B.6 (b) is a qualitative representation of the charges on the insulating tube, together with the polarization of the packthread and its connected ball. The packthread and the ball are conductors.

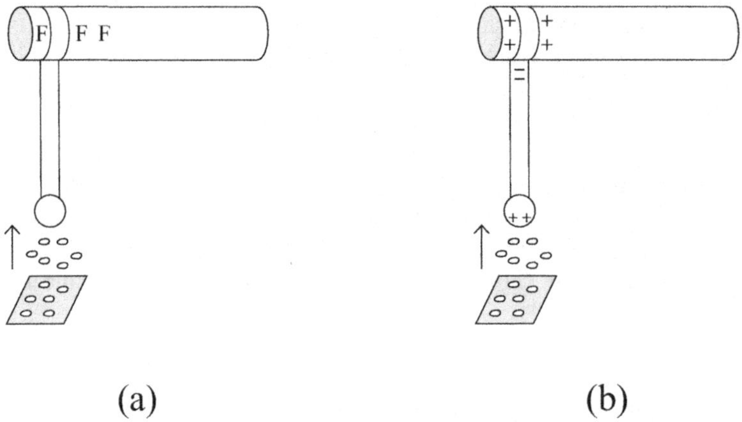

Figure B.6: (a) When Gray rubbed the glass tube, he observed the attraction of light brass leaf by bodies attached to the lower end of a packthread connected to the tube. The attracting body could even be metal. (b) Qualitative representation of the charges on the glass, together with the polarization of the packthread and connected ball.

By following these procedures, he was able to communicate the electricity of the rubbed tube to several bodies connected to it by strings or packthreads, such as coins, a fire-shovel, a copper tea-kettle empty or full of water, a silver pint pot, etc. In his words,[18] all these bodies "were strongly electrical, attracting the leaf-brass to the hight of several inches." Someone had finally succeeded in

[17] [Hei99, p. 246, Note].
[18] [Grah, p. 22].

making metals attract light bodies. No one had been able to obtain this effect in the 2,000 years since the discovery of electricity! As Heilbron said,[19] "[...] and so Gray succeeded at last in awakening their hidden electricity."

B.4 Gray Discovers Conductors and Insulators

He continued his research and with this technique electrified (or rather polarized) flint-stone, load-stone, several vegetable substances, etc. The brass-leaf could be attracted up to a height of 10 cm. After these experiments, he again worked with horizontal sticks attached to the glass tube. He set fishing rods 80 cm long into the tube. These rods also transmitted electricity, whether they were hollow or solid. Utilizing sticks and fishing rods, with a cork ball at the end, he was able to observe the effect even at 5.5 m distance. In May 1729 he continued his experiments, succeeding with a 7.3 m long wood pole connected to the glass tube. Even at this great length a cork ball connected at the end of the cane attracted a brass leaf when the tube was rubbed. He extended this to 9.7 m, including the tube. But once more the vibrations caused when he rubbed the tube disturbed the experiment. Again he decided to use a cork or ivory ball connected to the lower end of a packthread attached to the tube. When Gray rubbed the tube, he could cause the ball to attract brass leaf even with an 8 m long string, with Gray standing on the balcony. He then combined a long horizontal wood cane attached to the tube with a vertical thread connected at the other end of the cane, with an ivory ball at the lower end of the thread—like a huge fishing rod. Initially he worked with a 5.5 m long wood cane and a 10.3 m long thread. When he rubbed the glass tube, he observed the ivory ball attracting a brass leaf beneath it.

He then tried to increase the horizontal length by using only packthread. To do this, he made a loop at each end of a vertical packthread, its upper end hanging from a nail driven into a beam. The second packthread passed through the lower loop of the vertical packthread and was tied to the glass tube. The other end of this second packthread was tied to an ivory ball. If we follow this second packthread from the ivory ball to the glass tube, it will be vertical between the ball and the lower end of the first packthread, and it will be horizontal between this loop and the tube. Below the ivory ball he placed a brass leaf. In this case, when he rubbed the glass tube, Gray was not able to observe the slightest attraction of the brass leaf by the ivory ball (Figure B.7 (a)).

He then noted:[20]

> Upon this I concluded, that when the electrick vertue came [from the rubbed glass tube] to the loop that was suspended on the beam, it went up the same to the beam; so that none, or very little of it at

[19] [Hei99, p. 246].
[20] [Grah, p. 25].

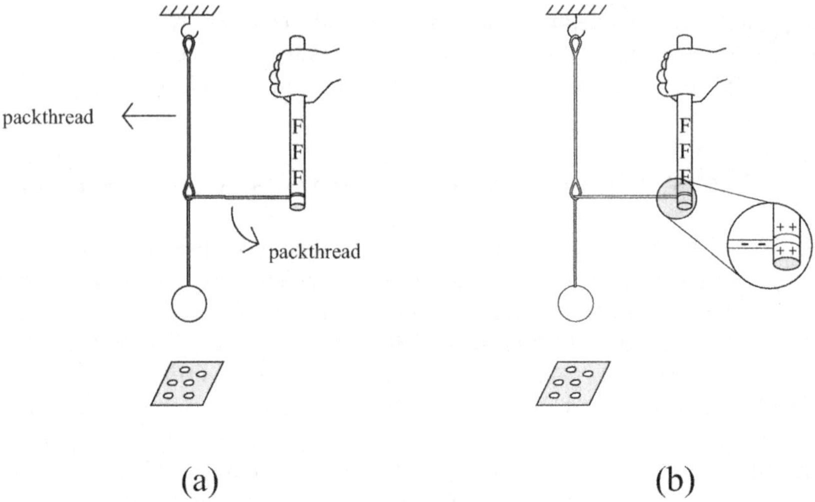

Figure B.7: (a) When the glass tube is rubbed, the ivory ball does not attract the brass leaf. In this situation, the packthread connected to the tube and to the ball is also connected to another packthread attached to the ceiling. (b) Qualitative representation of the charges in case (a).

> least, came down to the ball, which was afterwards verified, as will appear by the experiments that will be mentioned hereafter.

We now present the modern interpretation of this experiment. The rubbed glass tube initially polarizes the conducting packthread. But in this case the packthread connected to the tube is also connected to the Earth through another conducting packthread. This last packthread grounds the packthread connected to the glass, in analogy to what was seen in Figure 7.30. That is, the end of the packthread in contact with the glass acquires a charge of opposite sign to the glass. The other charges which were at the end of the ball in Figure B.6 (b) are now spread over the surface of the Earth due to the grounding. In Figure B.7 (b) we have a qualitative description of this experiment in terms of the charges spread over the glass and the packthread. In this case there is no net charge in the ball and the ball is not even polarized. For these reasons, the ball does not attract the metal leaves below it.

In July 1729 Gray decided to show these experiments to his friend Granville Wheler (1701-1770). Gray had a solid glass tube 28 cm long with 2 cm diameter. They attached the packthread to the tube, with a ball at the lower end of the packthread. Below the ball they placed the brass leaf. From a window they could make the ball attract the brass leaf by rubbing the glass tube with threads of 4.9 up to 10.4 m in length.

Gray continued his description of the experiments and then presented his greatest discovery:[21]

> As we had no greater heights here, Mr. Wheler was desirous to try whether we could not carry the electrick vertue horizontally. I then told him of the attempt I had made with that design, but without success, telling him the method and materials made use of, as mentioned above. He then proposed a silk line to support the line [of communication], by which the electrick vertue was to pass. I told him it might do better upon the account of its smallness [that is, Gray believed this could work better than in his original experiment due to the small thickness of the silk thread in comparison with the greater thickness of the packthread]; so that there would be less [electric] vertue carried from the line of communication, with which, together with the apt method Mr. Wheler contrived, and with the great pains he took himself, and the assistance of his servants, we succeeded far beyond our expectations.
>
> The first experiment was made in the matted Gallery July 2, 1729, about ten in the morning. About four feet [1.2 m] from the end of the Gallery there was a cross line that was fixed by its ends to each side of the Gallery by two nails; the middle part of the line was silk, the rest at each end packthread; then the line to which the ivory ball was hung, and by which the electrick vertue was to be conveyed to it from the tube, being eighty feet and a half [24.5 m] in length, was laid on the cross silk line, so as that the ball hung about nine feet [2.7 m] below it: Then the other end of the line [of communication] was by a loop suspended on the glass cane, and the leaf-brass held under the ball on a piece of white paper; when the tube being rubbed, the ball attracted the leaf-brass, and kept it suspended on it for some time.

A representation of this experiment can be found in Figure B.8 (a). A packthread is connected to a glass tube and to an ivory ball at the other end. Below the ball there are brass leaf. This packthread has an horizontal and a vertical portion. At the junction of these two portions, it is supported above a stretched silk thread. When Gray rubbed the glass, he observed the ball attracting the brass leaf below it. This attraction did not happen for the situation of Figure B.7. In this latter situation the string connected to the glass was suspended by another packthread attached to the ceiling.

In Figure B.8 (b) we have a qualitative representation of the charges in this experiment. In this case, the packthread is supported by an insulator, namely, by the silk thread. There is no grounding here. The situation is like that of Figure B.6 (b).

Here we have the fundamental discovery of conductors and insulators. For conductors we have cork, ivory ball, wood, packthread, metal wires, etc. For

[21][Grah, pp. 26-27].

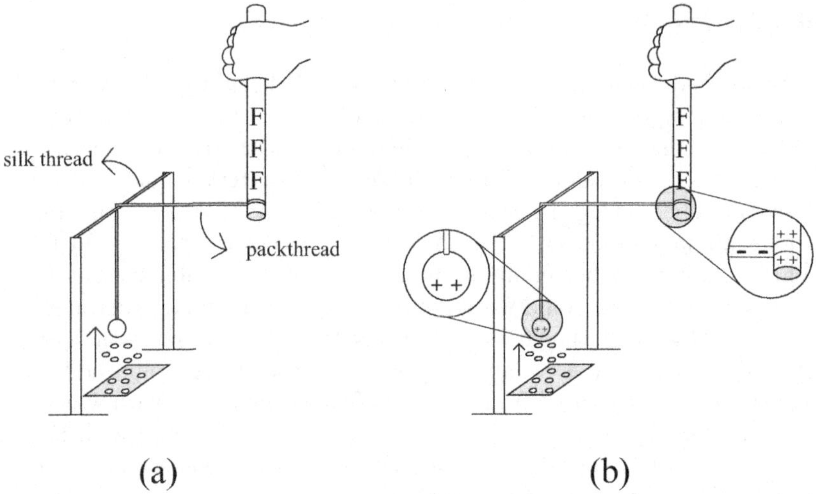

Figure B.8: (a) Gray observed an attraction on the brass leaf when he rubbed the glass tube. In this situation the packthread attached to the rubbed glass tube was supported by a silk thread. (b) Qualitative representation of the charges in case (a).

insulators we have the silk thread. Gray could communicate the electric virtue to the conductors through contact with a rubbed glass tube. The silk thread, on the other hand, did not allow the passage and dissipation of the electric virtue to the ground. During this article Gray described another insulator, namely, horse-hair fishing-lines.[22] In other articles from the same year Gray mentioned other insulators, namely, a cake of resin and warmed glass.[23] He used to make cakes of resin in order to support the bodies to which he wished to communicate the electric effluvium. In a paper of 1735 he also described cakes of beeswax, sulphur, and shell-lack.[24] All these materials he utilized as insulators or, in his words, as *electric bodies.*

Before continuing these quotations, it is important to remember the problem with old and new nomenclatures discussed in Section 8.1. That is, the substances Gilbert classified as *electric* are called *insulators* nowadays. The substances which were classified as *non-electric* are now called *conductors.*

A representation of this experiment appears in Figure B.9.[25]

This Figure shows Gray and his friend Wheler. Gray holds and rubs his 1 m long glass tube. Connected to the tube is a string with an ivory ball at the other end. The ball is close to ground, with small pieces of metal below it. The string connected to the glass tube is supported by other crossed lines. When these crossed lines are conductors, the ball does not attract the pieces of metal.

[22][Grah, p. 36].
[23][Grad, p. 228] and [Grag, pp. 399 and 406].
[24][Grae, pp. 18 and 20].
[25][Fig67, Vol. 1, Figure 227, p. 441], [Fig85, p. 321], [Bor], and [FM91, p. 88].

Figure B.9: Gray rubs his 1 m long flint glass tube with his bare hands. A packthread connected to the tube is supported by a silk thread. An ivory ball connected to the other end of the packthread attracts brass leaf beneath it.

On the other hand, when these crossed lines are made of an insulating material like silk, the ball attracts the pieces of metal below it when Gray rubs the glass tube.

An old representation of this crucial experiment by Gray is reproduced in Figure B.10.[26]

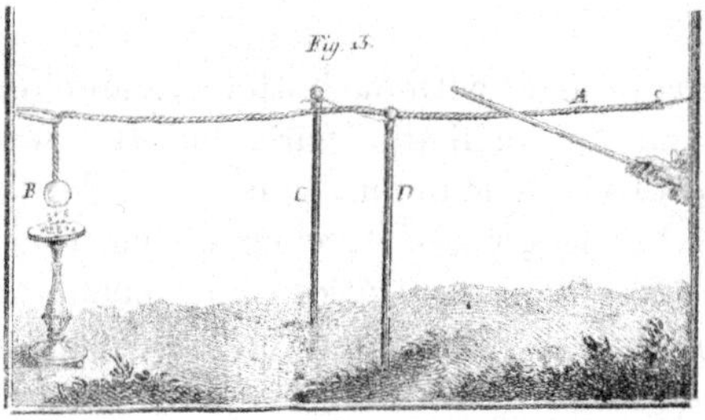

Figure B.10: An ivory ball attracts brass leaf when a rubbed glass tube touches a horizontal packthread, or when the tube is brought close to it, provided the packthread is supported by silk threads.

An interesting representation of Gray's experiment appears in Doppelmayr's book (Figure B.11).[27]

This is the second fundamental discovery described by Gray in this article, namely, the existence of conductors and insulators.

[26] [Nol53].
[27] [Dop74].

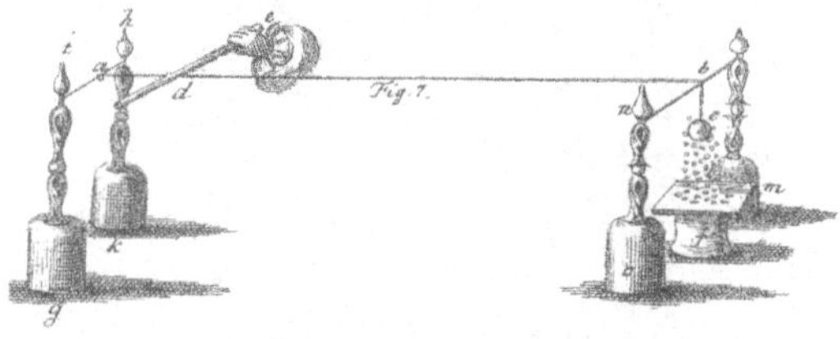

Figure B.11: A rubbed glass tube touches a horizontal rope and the ball attracts light bodies. The rope is supported by insulating strings.

B.5 Discovery that What makes a Body Behave as a Conductor or as an Insulator Depends upon Its Intrinsic Properties

After these experiments, Gray and Wheler were able to transmit the electric virtue 45 m horizontally by making turns in the conveyor line, that is, the packthread. They later reached 34 m along a straight horizontal line, together with 4 m vertically. On another day they reached 89 m with a horizontal thread making a few turns, always supported by silk lines. When they tried to increase this total length further, the silk line broke. It could not withstand the weight of the packthread and the vibrations occasioned by rubbing the glass tube.

Then came the third fundamental discovery of Gray, described in this article (our emphasys in italics):[28]

> Upon this, having brought with me both brass and iron wire, instead of the silk we put up small iron wire; but this was too weak to bear the weight of the [communication] line. We then took brass wire of a somewhat larger size [thickness] than that of iron. This supported our line of communication; but though the [glass] tube was well rubbed, yet there was not the least motion or attraction [of the brass leaf] given by the ball, neither with the great tube [of glass 1 m long], which we made use of when we found the small solid cane [of glass 28 cm long] to be ineffectual: *By which we were now convinced, that the success we had before, depended upon the lines that supported the line of communication, being silk, and not upon their being small [thin], as before trial I imagined it might be;* the same effect happened here as it did when the line that is to convey the electrick vertue is supported by packthread; viz. that when the [electric] effluvia come to the wire or packthread that supports the [communication] line, it passes by them to the timber, to which each end of them is fixed, and so goes no farther forward in the line that is to carry it to the ivory ball.

[28][Grah, pp. 28-29].

Gray had already discovered how to communicate the electric virtue to wood, metals, and several other substances. He had also discovered that a silk thread prevented the loss of electricity through it. However, he initially believed that this insulating property was due to the small thickness of the silk thread, in comparison with the larger thickness of a packthread. With the present experiment, on the other hand, he discovered that two thin crossed lines of almost the same thicknesses, one metal and the other made of silk, exhibited completely different behaviour. While the metal wire (or the packthread) allowed the passage of the electric virtue to the ground, the silk thread did not allow the flow of electricity through it. This meant that in his experiments it was essentially the kind of material which defined or characterized its property. The sizes or thicknesses of the materials were not so relevant in his experiments in determining whether they would behave as a conductor or as an insulator, contrarily to what he had originally thought. This was his third fundamental discovery.

B.6 Discovery that Electrification by Communication Happens at a Distance

They went on with their experiments, transmitting electricity up to 203 m, with the packthread making eight returns supported by silk threads. A representation of this experiment appears in Figures B.12 and B.13.[29]

Figure B.12: A representation of Gray's experiment.

In straight line they extended the experiment to 198 m, and later on 233 m. In subsequent experiments they varied the bodies which were suspended at

[29] [GS89], [BWb], and [Dop74].

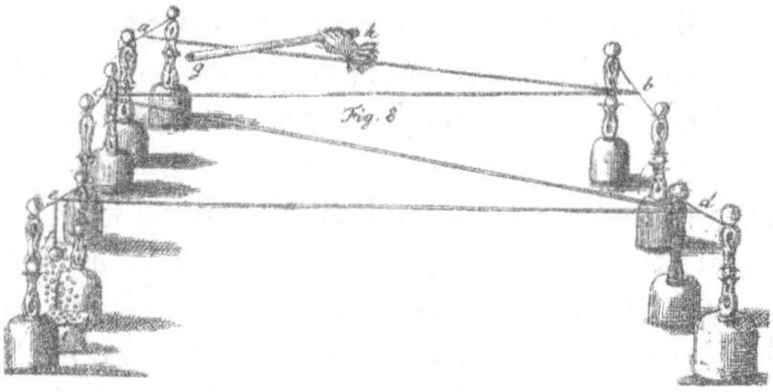

Figure B.13: A ball attracting light bodies when supported by insulating strings. The rubbed glass approaches the other end of the conducting cord.

the free end of the packthread. Instead of an ivory ball, they used a large 8 m^2 map of the world, an umbrella, and a loadstone with a metal key hung by one of its arming irons. All these materials attracted the brass leaf when they rubbed the glass tube. After this experiment, they suspended three bodies in different locations along the communication line. All of them attracted brass leaf simultaneously when the glass was rubbed. They also suspended a live chick by the legs and observed that its breast became strongly attracting.

At the end of the paper he presented other experiments showing that he could transmit the electric virtue up to a distance of 270 m.

Then came the fourth important discovery described by Gray in this paper. He showed that he could transmit the electric virtue along the communication line simply by bringing a rubbed glass tube near to one of its ends, without contact between the glass tube and the packthread:[30]

> *At Mr. Godfrey's I made the following experiments; showing that the electrick vertue may be carried from the tube, without touching the line of communication, by only being held near it.*
>
> The first of these experiments was made the 5th of August, 1729. [...]
>
> I took a piece of a hair-line, such as linnen cloaths are dried on, of about eleven feet [3.3 m] in length; which, by a loop at the upper end of it, was suspended on a nail, that was drove into one of the rafters in the garret, and had at its lower end a leaden weight of fourteen pounds [6.4 kg] hung to it by an iron ring: then the leaf-brass was laid under the weight, and the tube rubbed, and being held near the line without touching it, the lead-weight attracted and repelled the leaf-brass for several times together, in the hight of at least three, if not four inches [10 cm]. If the tube was held three or four feet [1.2 m] above the weight, there would be an attraction; but if it were held higher up, so as to be near the rafter where the weight was hung by the hair-line, there would be no attraction.

[30] [Grah, pp. 33-34].

A representation of this experiment can be found in Figure B.14 (a).

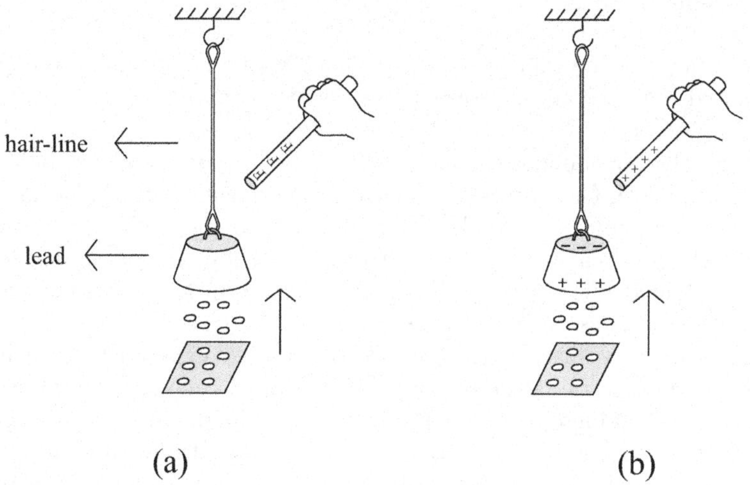

Figure B.14: (a) Gray was able to make the metal weight attract light bodies by simply bringing the rubbed glass near the weight, without touching it. (b) Qualitative representation of the charges in case (a).

Our present day understanding of this experiment is that the rubbed glass tube electrically polarizes the lead weight. The portion of the weight which is closer to the glass becomes electrified with a charge of opposite sign to the glass, while the farthest portion of the lead becomes electrified with a charge of the same sign as that of the glass. The brass leaf is then attracted essentially by the lower portion of the polarized lead. In a certain sense this is analogous to Experiment 7.9.

A qualitative representation of the charges in this experiment appears in Figure B.14 (b).

Gray also utilized *hair lines* to suspend bodies in other experiments. It is not clear exactly what they were made of. In any event, they functioned as insulators. In another famous experiment described in this article of 1731, Gray suspended a 21 kg boy in a horizontal position,[31] "*by two hair-lines, such as cloaths are dried on.*" He then brought a rubbed glass tube close to the boy's feet, without touching them, and observed that the boy's face attracted brass leaf placed below him.

Du Fay repeated the experiment with the boy in 1733. When he used common strings ("cordes ordinaires") he did not obtain any attraction. However, when he replaced the common strings with silk cords ("cordons de soye"), he obtained the same attractions as Gray had obtained.[32] The common strings are usually conductors. This shows that Gray's *hair lines* are insulators, since only when we use insulators are the experiments with the boy successful.

[31] [Grah, p. 39].
[32] [DF33d, pp. 250-251].

In a paper from 1735 Gray performed some similar experiments. He began by mentioning that[33]

> As I had not any silk lines by me strong enough to bear the boy, I caused him to stand on some of the electric bodies.

That is, the boy stood on some insulators, as we say nowadays. In the next page of this article, Gray described another experiments which he performed at Mr. Wheler's home: "Mr. Wheler, soon after my coming to him, procured silk lines strong enough to bear the weight of his footboy, a good stout lad; then having suspended him upon the lines, [...]." From all this it is very probable that Gray's *hair lines* were made of silk.

In the sequel to this paper, Gray described other experiments in which he transmitted the electric virtue to conductors simply by moving the rubbed glass tube near the conductors, without touching the conductors. Utilizing wood hoops 66 cm and 91 cm in diameter, suspended by insulating strings, he observed that the electric effluvium might be transported along the circumference of these hoops. It was also able to pass from one hoop to another hoop in contact with the first hoop (Figure B.15).[34] He could also transmit the electric virtue to many fruits and vegetables.

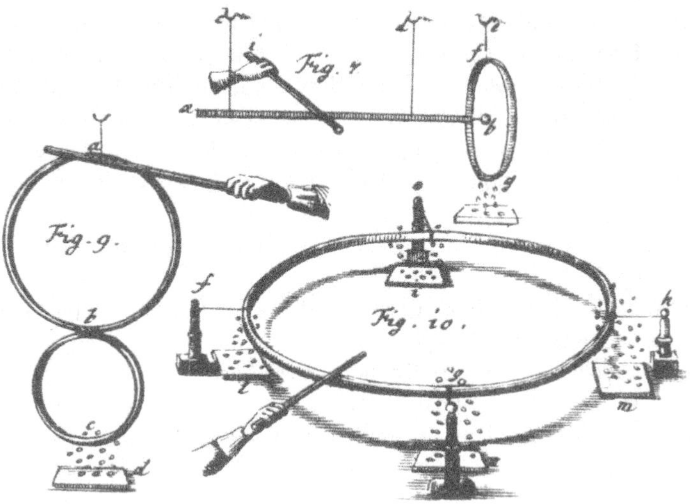

Figure B.15: Representation of Gray's experiments with wood hoops supported by insulating strings.

Gray was even able to make a soap bubble attract light bodies:[35]

[33][Grae, p. 17].
[34][Dop74] and [Hei99, p. 249].
[35][Grah, pp. 38-39].

> March the 23d [of 1730], I dissolved soap in the Thames-water, then I suspended a tobacco-pipe by a hair-line [probably of silk or a horse-hair], so as that it hung nearly horizontal, with the mouth of the bowl downwards; then having dipped it in the soap-liquor, and blown a bubble, the leaf-brass laid on a stand under it, the [glass] tube being rubbed, the brass was attracted by the bubble, when the tube was held near the hair-line. Then I repeated the experiment with another bubble, holding the tube near the little end of the pipe, and the attraction was now much greater, the leaf-brass being attracted to the hight of near two inches [5 cm].

This experiment, represented in Figure B.16 (a), illustrates once more that fresh water behaves as a conductor. Gray had already transmitted the capacity of attracting light bodies to many conductors, such as metals, wood, etc.

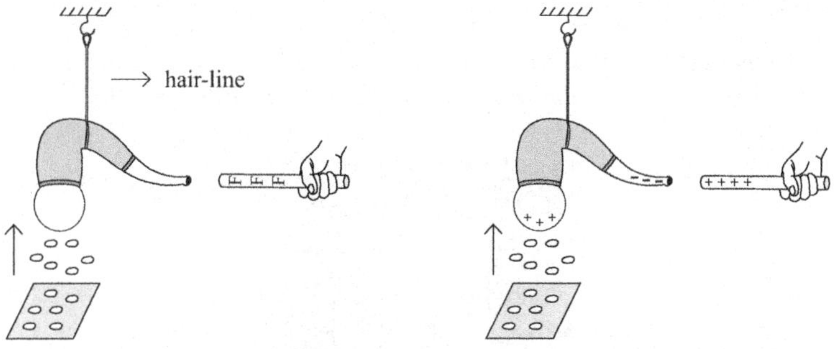

(a) (b)

Figure B.16: (a) A soap bubble attracting brass leaves when the rubbed glass tube is brought near the pipe suspended by a hair-line, an insulator. (b) Qualitative representation of the charges in case (a).

Figure B.16 (b) presents a qualitative representation of the charges in this experiment.

B.7 The Experiment with the Suspended Boy

In this work of 1731 Gray described several experiments in which he suspended a boy in a horizontal position by hair lines, probably made of silk.[36] For instance, with the face of the boy pointing downward, Gray held the rubbed glass tube near his feet, without touching them, and observed brass leaves being attracted to the boy's face, lifting up to 30 cm. An old representation of this experiment can be found in Figure B.17.[37]

[36] [Grah, pp. 39-41].
[37] [Dop74].

Figure B.17: A boy is suspended by insulating lines. A rubbed glass tube is moved near his legs. The hands and face of the boy attract light bodies.

This experiment became very famous. It was used by Nollet in the frontispiece of his book *Essai sur l'Électricité des Corps* (Figure B.18).[38]

Figure B.18: Representation of Gray's famous experiment in Nollet's book, [Nol53]. In this figure it can be seen that the glass tube does not need to touch the boy.

Some of these experiments by Gray were repeated and extended by Du Fay. For instance, in his third memoir he has an experiment that is illustrated here in Figure B.19.[39]

[38] [Nol53].

[39] [DF33d, pp. 248-249] and [RR57, p. 584].

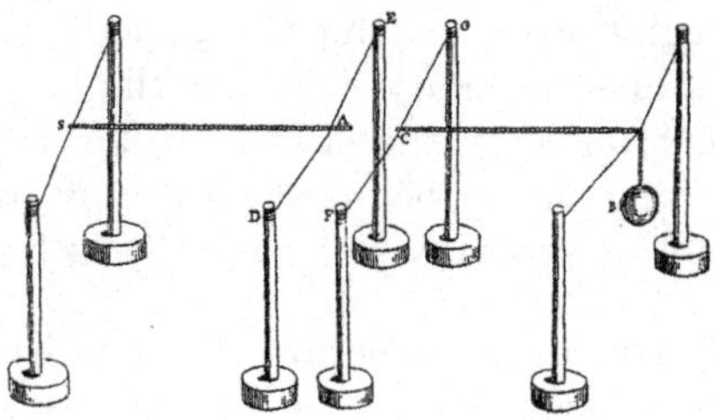

Figure B.19: Illustration of Du Fay's experiment analogous to some earlier experiments due to Gray.

Du Fay described this experiment as follows:[40,41]

> I took two pieces of a [conducting] line, with the thickness of a finger, in which the first one SA, had a length of 6 feet [1.8 m], and the other CB had 8 [feet long, that is, 2.4 m], I fixed each one of them by one end to two silk cross cords, DE and FG, at right angles to them, and which were disposed in such a way that we could approach or move away parallelly these cross cords from one another, so that we could fix them at the chosen distance [from one another]. At the end B of the 8 feet line a wood ball was suspended, and the remotest end of the 6 feet line was fixed to a third silk cross cord at S in order to suspend it in air. Then, by presenting the rubbed [glass] tube to the S end of the cord SA, after having separated the two lines one inch [2.54 cm] from one another, the electricity was so sensible at the ball as if the line had been continuous [observing that the ball attracted small metal leaves placed near it], at [a separation of] 3 inches [7.5 cm] it [the electricity] was even more [sensible], at 6 inches [15 cm] a little less, and at 1 foot [30 cm] much less, more or less like at a distance of 1256 feet of a continuous line [377 m as Du Fay had experienced before]. Therefore, the electric substance

[40] [DF33d, pp. 248-249].

[41] J'ai pris deux morceaux d'un cordon de fil, gros comme le doigt, dont le premier SA, avoit 6 pieds de long, & l'autre CB, en avoit 8, je les ai assujettis chacun par un bout à deux brides de soye DE, & FG, qui les coupoient à angles droits, & qui étoient disposées de sorte qu'approchant ou éloignant parallelement ces brides l'une de l'autre, les deux bouts des deux cordons s'éloignoient ou s'approchoient l'un de l'autre, de maniére qu'on pouvoit les fixer à la distance qu'on souhaitoit. Au bout B du cordon de 8 pieds étoit suspenduë une boule de bois, & le bout le plus éloigné du cordon de 6 pieds étoit fixé à une troisiéme bride de soye en S pour la soûtenir en l'air; présentant ensuite le tube frotté au bout S du cordon SA, après avoir éloigné les deux cordons d'un pouce l'un de l'autre, l'électricité étoit aussi sensible dans la boule que si le cordon eût été continu, à 3 pouces elle l'étoit encore beaucoup, à 6 pouces un peu moins, & à 1 pied beaucoup moins, & à peu-près comme à la distance de 1256 pieds de corde continuë; la matiére électrique coule donc librement dans l'air, sans être fixée par aucun corps. Cette expérience prouve combien il est nécessaire que la corde dont on se sert pour transmettre au loin l'électricité, soit isolée, ou ne soit soûtenuë que de corps les moins propres qu'il est possible à se charger eux-mêmes de l'électricité.

> flowed freely through the air, without being fixed by any body. This experiment proves how necessary it is that the [conducting] string utilized to transmit far away the electricity, be *insulated*, that is, [the conducting string should be] suspended only by bodies which are the least possible appropriate to charge themselves of electricity.

The modern description of this experiment is not based upon an electric substance flowing freely through the air, as Du Fay imagined. Instead of this, it is considered that the main phenomenon in this experiment is the electric polarization of conductors, as in Figure B.20. That is, the rubbed glass tube electrically polarizes the conducting rope SA when brought near it. The charges at the end A of this rope electrically polarize another conductor CB. This conductor CB is composed of a second rope connected to a wooden ball. Both polarizations take place despite the air gap between A and C, varying from one inch, 2.5 cm, to one foot, 30 cm. The charges accumulated at the lower end of the wooden ball, which have the same sign as the charges on the rubbed glass tube, attract light bodies nearby.

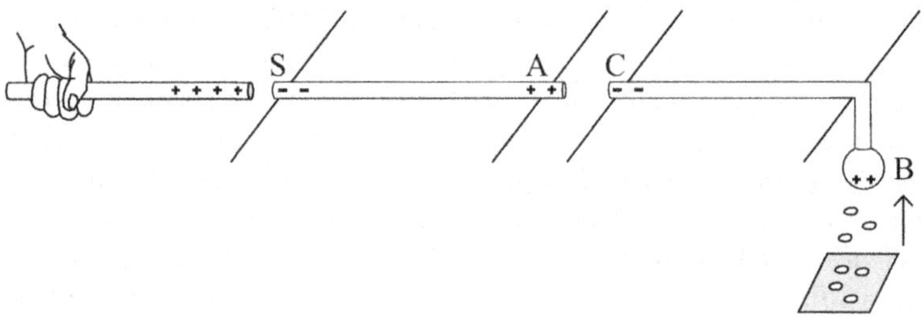

Figure B.20: Illustration of polarization in Du Fay's experiment (Figure B.19). Conductors SA and CB are supported above insulating silk threads.

B.8 Discovery that Free Charges are Distributed over the Surface of Conductors

In the sequel to Gray's 1731 paper, he described another fundamental discovery, namely (our emphasis in italics):[42]

> *Some time after, at Mr. Wheler's, we made the following experiment, in order to try whether the electrick attraction be proportional to the quantity of matter in bodies.*
>
> There were made two cubes of oak, of about six inches square [15 cm^2], the one solid, the other hollow: These were suspended by two hair-lines, nearly after the same manner as in the experiment above-mentioned; the distance of the cubes from each other, was by

[42][Grah, p. 35].

estimation, about fourteen or fifteen feet [4.6 m]; the line of communication being tied to each hair-line, and the leaf-brass placed under the cubes, the [glass] tube was rubbed and held over the middle of the [communication] line, and as near as could be guessed, at equal distances from the cubes, when both of them attracted and repelled the leaf-brass at the same time, and to the same hight; *so that it seemed to be no more attraction in the solid than in the hollow cube; yet I am apt to think that the electrick effluvia pass through all the interior parts of the solid cube, though no part but the surface attracts*; for from several experiments it appears, that if any solid body touches that which attracts, its attraction ceases till that body be removed, and the other be again excited by the tube.

A representation of this experiment can be found in Figure B.21.

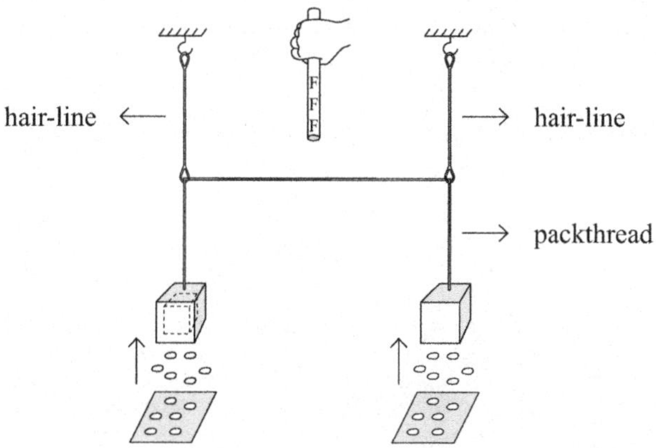

Figure B.21: A hollow cube and a solid cube attract with the same force.

This experiment describes two extremely important discoveries. The first is that in electrostatics the free charges, or the excess charges upon conductors (as the conducting wooden cubes in this experiment), are distributed over the surface of the conductors, and not throughout their volumes. Sometimes this fundamental property of conductors in electrostatic equilibrium is attributed to Michael Faraday (1791-1867). This experiment shows that this fact was already known by Gray.[43]

The second discovery, expressed in the last sentence quoted above, is the fact that an electrified conductor is discharged when it touches another conductor connected to the ground, that is, when the electrified conductor is grounded. Gray seems to be referring here to his experiments of 1708. See Sections 4.2 and 4.5.

43[CM79, p. 396] and [Hei99, pp. 248-249].

B.9 Discovery of the Power of Points

Initially Gray laid the brass leaf on a stand, which was a round board 30 cm in diameter, with white paper pasted on it, supported on a 30 cm high pedestal. In the sequel to his historic paper, Gray even described another very important discovery, namely:[44]

> In these experiments, besides the large stand above-mentioned, I made use of two small ones, which, as I found them very useful, it may not be improper to describe them. The tops of them were three inches [7.6 cm] diameter; they were supported by a column of about a foot [30 cm] in hight, their bases of about four inches and a half [11.4 cm]: They were turned of *Lignum vitae* [a type of wood], their tops and bases made to skrew on for convenience of carriage. Upon the tops were pasted white paper. When the leaf-brass is laid on any of these stands, I find it is attracted to a much greater hight than when laid on a table, and at least three times higher than when laid on the floor of a room.

A representation of this experiment can be found in Figure B.22.

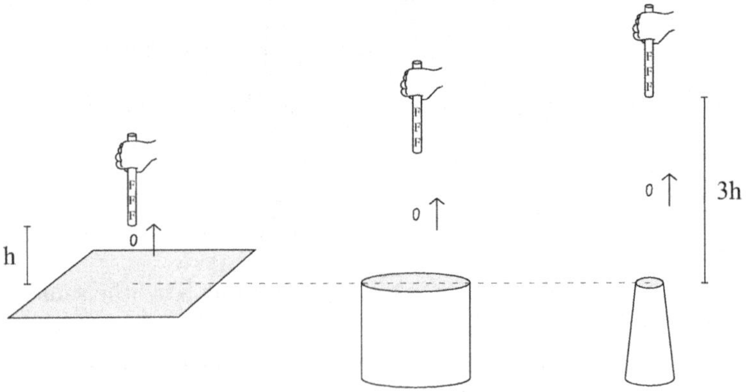

(a) (b) (c)

Figure B.22: (a) A small leaf-brass is attracted to a height h from its initial position on the floor. (b) It rises higher when laid on a table or on a conducting cylinder 30 cm in diameter. (c) When it is on a conical conductor, it is attracted three times higher than when it is on the floor.

This is one of the first known descriptions of the power of points. That is, the electric force is stronger around sharp and pointed regions of conductors than around flat surfaces.

[44] [Grah, p. 42].

In Section 4.10 we analyzed the behaviour of the arrows attached to an electric pendulum. They pointed toward the attracting rubbed straw before coming into contact with it. After contact, they pointed away from the straw. This behaviour is related with the power of points discovered by Gray.

B.10 Conclusion

No doubt this is one of the most important papers in the whole history of electricity. The number of fundamental discoveries made by a simple retired dyer who never studied at a university is truly impressive. At this time in his life he was 63 years old. We consider Gray's main contribution his discovery of conductors and insulators. He described some of their main properties. This allowed a control of electric phenomena, paving the way for a series of new discoveries which were made soon afterwards by Gray and by other scientists. In his other papers Gray described many new extremely important discoveries in electricity, but we will not discuss them here.

Du Fay followed the works of Gray closely, and was strongly influenced by his papers. Du Fay's discoveries of electric repulsion, the *ACR* mechanism, and the two kinds of electricity, were made after he decided to reproduce and explore the many discoveries Gray had made earlier. For instance, in one of his most important works, Du Fay said the following:[45]

> I beseech your Grace to communicate it [this letter] to the *Royal Society*, and in particular to Mr. *Gray*, who works on this subject with so much application and success, and to whom I acknowledge my self indebted for the discoveries I have made, as well as for those I may possibly make hereafter, since 'tis from his writings that I took the resolution of applying my self to this kind of experiments.

We began this book with an account of the amber effect, an experiment known at least since Plato's time, the IVth century B.C. We concluded it with a description of the works of a retired dyer whose discoveries gave us a great advance in our understanding of nature and in the technical field of electricity. The ways science developed are fascinating indeed!

[45][DF, pp. 265-266] and [BC07, p. 643].

Bibliography

[Aep79] F. U. T. Aepinus. *Aepinus's Essay on the Theory of Electricity and Magnetism.* Princeton University Press, Princeton, 1979. Translation by P. J. Connor. Introductory monograph and notes by R. W. Home.

[AH07] A. K. T. Assis and J. A. Hernandes. *The Electric Force of a Current: Weber and the Surface Charges of Resistive Conductors Carrying Steady Currents.* Apeiron, Montreal, 2007. ISBN: 978-0-9732911-5-5. Available at: www.ifi.unicamp.br/~assis.

[AH09] A. K. T. Assis and J. A. Hernandes. *A Força Elétrica de uma Corrente: Weber e as Cargas Superficiais de Condutores Resistivos com Correntes Constantes*, volume 73 of *Acadêmica.* Edusp and Edufal, São Paulo and Maceió, 2009. ISBNs: 978-85-314-1123-6 and 978-85-7177-431-5.

[Amp22] A.-M. Ampère. Expériences relatives à de nouveaux phénomènes électro-dynamiques. *Annales de Chimie et de Physique*, 20:60–74, 1822.

[Ari52a] Aristotle. *Metaphysics*, In: The Works of Aristotle, Vol. 1. In *Great Books of the Western World,* R. M. Hutchins, Editor in Chief, Vol. 8, pages 495–626, Chicago, 1952. Encyclopaedia Britannica. Translated by W. D. Ross.

[Ari52b] Aristotle. *Meteorology*, In: The Works of Aristotle, Vol. 1. In *Great Books of the Western World,* R. M. Hutchins, Editor in Chief, Vol. 8, pages 443–494, Chicago, 1952. Encyclopaedia Britannica. Translated by E. W. Webster.

[Ari52c] Aristotle. *On the Soul*, In: The Works of Aristotle, Vol. 1. In *Great Books of the Western World,* R. M. Hutchins, Editor in Chief, Vol. 8, pages 627–668, Chicago, 1952. Encyclopaedia Britannica. Translated by J. A. Smith.

[Ass08a] A. K. T. Assis. *Archimedes, the Center of Gravity, and the First Law of Mechanics.* Apeiron, Montreal, 2008. ISBN: 978-0-9732911-6-2. Available at: www.ifi.unicamp.br/~assis.

[Ass08b] A. K. T. Assis. *Arquimedes, o Centro de Gravidade e a Lei da Alavanca.* Apeiron, Montreal, 2008. ISBN: 978-0-9732911-7-9. Available at: www.ifi.unicamp.br/~assis.

[BC07] S. L. B. Boss and J. J. Caluzi. Os conceitos de eletricidade vítrea e eletricidade resinosa segundo Du Fay. *Revista Brasileira de Ensino de Física*, 29:635–644, 2007.

[BC09] S. L. B. Boss and J. J. Caluzi. Uma breve biografia de Stephen Gray (1666-1736). Accepted for publication in Revista Brasileira de Ensino de Física, 2009.

[Ben86] A. Bennet. Description of a new electrometer. *Philosophical Transactions*, 76:26–34, 1786.

[Ben98] P. Benjamin. *A History of Electricity (The Intellectual Rise in Electricity) from Antiquity to the Days of Benjamin Franklin.* Wiley, New York, 1898.

[Blo82] C. Blondel. *A.-M. Ampère et la Création de l'Électrodynamique (1820-1827).* Bibliothèque Nationale, Paris, 1982.

[Bor] G. Borvon. Pourquoi deux espèces d'électricité? Pourquoi deux sens du courant électrique? L'histoire de l'électricité nous aide à comprendre. Available at: www.ampere.cnrs.fr/parcourspedagogique.

[Boy00] R. Boyle. Experiments and notes about the mechanical origine or production of electricity. In M. Hunter and E. B. Davis, editors, *The Works of Robert Boyle*, Vol. 8, pages 509–523. Pickering & Chatto, London, 2000. Work originally published in 1675.

[BWa] C. Blondel and B. Wolff. Que dit l'article ELECTRICITE de l'Encyclopédie? Available at: www.ampere.cnrs.fr/parcourspedagogique.

[BWb] C. Blondel and B. Wolff. Teinturiers et tubes de verre: Gray et Dufay. Available at: www.ampere.cnrs.fr/parcourspedagogique.

[BWc] C. Blondel and B. Wolff. Un tournant dans l'histoire de l'électricité: la mathématisation. Available at: www.ampere.cnrs.fr/parcourspedagogique.

[CA08] J. Camillo and A. K. T. Assis. Construção de um gerador eletrostático gotejante: chuva elétrica de Kelvin. *A Física na Escola*, 9:29–32, 2008.

[Can53] J. Canton. Electrical experiments, with an attempt to account for their several phaenomena; together with observations on thunderclouds. *Philosophical Transactions*, 48:350–358, 1753.

[Can54] J. Canton. A letter to the Right Honourable the Earl of Macclesfield, President of the Royal Society, concerning some new electrical experiments. *Philosophical Transactions*, 54:780–785, 1754.

[CC00] D. H. Clark and S. P. H. Clark. *Newton's Tyranny: The Suppressed Scientific Discoveries of Stephen Gray and John Flamsteed.* Freeman, New York, 2000.

[Cha09] J. P. M. d. C. Chaib. *Análise do Significado e da Evolução do Conceito de Força de Ampère, juntamente com a Tradução Comentada de sua Principal Obra sobre Eletrodinâmica.* PhD thesis, Universidade Estadual de Campinas—UNICAMP, Campinas, Brazil, 2009. Supervisor: A. K. T. Assis. Available at http://webbif.ifi.unicamp.br/teses/ and at www.ifi.unicamp.br/~assis.

[Chi54] R. A. Chipman. An unpublished letter of Stephen Gray on electrical experiments, 1707-1708. *Isis*, 45:33–40, 1954.

[CM79] D. H. Clark and L. Murdin. The enigma of Stephen Gray astronomer and scientist (1666-1736). *Vistas in Astronomy*, 23:351–404, 1979.

[Desa] J. T. Desaguliers. Experiments made before the Royal Society, Feb. 2. 1737-8. *Philosophical Transactions*, 41:193–199, 1739–1741.

[Desb] J. T. Desaguliers. Several electrical experiments, made at various times, before the Royal Society. *Philosophical Transactions*, 41:661–667, 1739–1741.

[Desc] J. T. Desaguliers. Some thoughts and experiments concerning electricity. *Philosophical Transactions*, 41:186–193, 1739–1741.

[DF] C. F. d. C. Du Fay. A letter from Mons. Du Fay, F. R. S. and of the Royal Academy of Sciences at Paris, to His Grace Charles Duke of Richmond and Lenox, concerning electricity. Translated from the French by T. S. M D. *Philosophical Transactions*, 38:258–266, 1733–4.

[DF33a] C. F. d. C. Du Fay. Premier mémoire sur l'électricité. Histoire de l'électricité. *Mémoires de l'Académie Royale des Sciences*, pages 23–35, 1733.

[DF33b] C. F. d. C. Du Fay. Quatriéme mémoire sur l'électricité. De l'attraction et répulsion des corps électriques. *Mémoires de l'Académie Royale des Sciences*, pages 457–476, 1733.

[DF33c] C. F. d. C. Du Fay. Second mémoire sur l'électricité. Quels sont les corps qui sont susceptibles d'électricité. *Mémoires de l'Académie Royale des Sciences*, pages 73–84, 1733.

[DF33d] C. F. d. C. Du Fay. Troisiéme mémoire sur l'électricité. Des corps qui sont le plus vivement attirés par les matiéres électriques, et de ceux qui sont les plus propres à transmettre l'électricité. *Mémoires de l'Académie Royale des Sciences*, pages 233–254, 1733.

[DF34a] C. F. d. C. Du Fay. Cinquiéme mémoire sur l'électricité. Oú l'on rend compte des nouvelles découvertes sur cette matiére, faites depuis peu par M. Gray; et où l'on examine quelles sont les circonstances qui peuvent apporter quelque changement à l'électricité pour l'augmentation ou la diminution de la force, comme la tempèrature de l'air, le vuide, l'air comprimé, etc. *Mémoires de l'Académie Royale des Sciences*, pages 341–361, 1734.

[DF34b] C. F. d. C. Du Fay. Sixiéme mémoire sur l'électricité. Oú l'on examine quel rapport il y a entre l'électricité, et la faculté de rendre de la lumiére, qui est commune à la plûpart des corps électriques, et ce qu'on peut inférer de ce rapport. *Mémoires de l'Académie Royale des Sciences*, pages 503–526, 1734.

[DF37a] C. F. d. C. Du Fay. Huitiéme mémoire sur l'électricité. *Mémoires de l'Académie Royale des Sciences*, pages 307–325, 1737.

[DF37b] C. F. d. C. Du Fay. Septiéme mémoire sur l'électricité. Contenant quelques additions aux mémoires précédents. *Mémoires de l'Académie Royale des Sciences*, pages 86–100, 1737.

[Dop74] J. G. Doppelmayr. *Neu-entdeckte Phaenomena von bewunderswürdigen Würkungen der Natur.* Nurenburg, 1774.

[Fera] N. Ferreira. *Mecânica.* Instituto de Física, USP, São Paulo. Projeto RIPE — Rede de Instrumentação para o Ensino.

[Ferb] N. Ferreira. *Eletrostática*, Vol. 1. Instituto de Física, USP, São Paulo. Projeto RIPE — Rede de Instrumentação para o Ensino.

[Ferc] N. Ferreira. *Eletrostática*, Vol. 2. Instituto de Física, USP, São Paulo. Projeto RIPE — Rede de Instrumentação para o Ensino.

[Ferd] N. C. Ferreira. Construa sua própria bússola! Available at: cienciahoje.uol.com.br/controlPanel/materia/view/2888.

[Fer78] N. C. Ferreira. Proposta de Laboratório para a Escola Brasileira— Um Ensaio sobre a Instrumentalização no Ensino Médio de Física. Master's thesis, Universidade de São Paulo, São Paulo, 1978.

[Fer01a] N. C. Ferreira. Acende aqui, apaga ali. *Ciência Hoje na Escola*, 12:65–67, 2001.

[Fer01b] N. C. Ferreira. Faça como Gilbert: construa uma bússola de declinação. *Ciência Hoje na Escola*, 12:21–22, 2001.

[Fer01c] N. C. Ferreira. Magnetismo e eletricidade. *Ciência Hoje na Escola*, 12:14–17, 2001.

[Fer01d] N. C. Ferreira. O versorium. *Ciência Hoje na Escola*, 12:18–20, 2001.

[Fer06] N. Ferreira. *Equilíbrio*. Projeto RIPE, Instituto de Física, USP, São Paulo, 2006. Available at: www.ludoteca.if.usp.br.

[Fig67] L. Figuier. *Ler Merveilles de la Science ou Description Populaire des Inventions Modernes*, volume 1. Jouvet et Cie., Paris, 1867.

[Fig85] L. G. Figuier. *Les Merveilles de l'Électricité*. Association pour l'Histoire de l'Électricité en France, Paris, 1985. Textes choisis présentés par Fabienne Cardot.

[FM91] N. Ferreira and J.-P. Maury. *Plus et Moins, les Charges Électriques. Qu'est-ce que c'est?* Ophrys, Paris, 1991.

[Fra55] G. Fracastoro. *De sympathia & antipathia rerum*. In: *Opera omnia in unum proxime post ilius mortem collecta*. Apud luntas, Venice, 1555.

[Gas91] A. Gaspar. Motor de ímã móvel. *Caderno Catarinense de Ensino de Física*, 8:188–193, 1991.

[Gas96] A. Gaspar. *História da Eletricidade*. Ática, São Paulo, 1996.

[Gas03] A. Gaspar. *Experiências de Ciências para o Ensino Fundamental*. Ática, São Paulo, 2003.

[GG90] I. Grattan-Guinness. *Convolutions in French Mathematics, 1800-1840*, volume 2. Birkhäuser, Basel, 1990.

[GG91] I. Grattan-Guinness. Lines of mathematical thought in the electrodynamics of Ampère. *Physis*, 28:115–129, 1991.

[Gil00] W. Gilbert. *De magnete, magneticisque corporibus, et de magno magnete tellure; physiologia nova*. London, 1600.

[Gil00] W. Gilbert. *On the Magnet, Magnetick Bodies also, and on the Great Magnet the Earth; a New Physiology, Demonstrated by Many Arguments & Experiments*. Chiswick Press, London, 1900. Translated by S. P. Thompson.

[Gil78] W. Gilbert. *On the Loadstone and Magnetic Bodies and on the Great Magnet the Earth*, volume 28, pp. 1-121 of *Great Books of the Western World*. Encyclopaedia Britannica, Chicago, 1978. Translated by P. F. Mottelay.

[Gli33] M. Gliozzi. L'elettrologia nel secolo XVII. *Periodico di Matematiche*, 13:1–14, 1933.

[Graa] S. Gray. An account of some electrical experiments intended to be communicated to the Royal Society by Mr. Stephen Gray, F. R. S. taken from his mouth by Cromwell Mortimer, M. D. R. S. Secr. on Feb. 14, 1735-6. Being the day before he died. *Philosophical Transactions*, 39:400–403, 1735–1736.

[Grab] S. Gray. An account of some new electrical experiments. *Philosophical Transactions*, 31:104–107, 1720–1.

[Grac] S. Gray. Concerning the revolutions which small pendulous bodies will, by electricity, make round larger ones from West to East as the planets do round the Sun. *Philosophical Transactions*, 39:220, 1735–1736.

[Grad] S. Gray. The electricity of water. *Philosophical Transactions*, 37:227–230 (addenda in page 260), 1731–2.

[Grae] S. Gray. Experiments and observations upon the light that is produced by communicating electrical attraction to animate or inanimate bodies, together with some of its most surprising effects. *Philosophical Transactions*, 39:16–24, 1735–6.

[Graf] S. Gray. Farther account of his experiments concerning electricity. *Philosophical Transactions*, 37:285–291, 1731–2.

[Grag] S. Gray. Farther accounts of his experiments concerning electricity. *Philosophical Transactions*, 37:397–407, 1731–2.

[Grah] S. Gray. Several experiments concerning electricity. *Philosophical Transactions*, 37:18–44, 1731–2.

[Grai] S. Gray. Some experiments relating to electricity. *Philosophical Transactions*, 39:166–170, 1735–6.

[GS89] G. Gaudenzi and R. Satolli. *Jean-Paul Marat: Scienziato e Rivoluzionario*. Mursia, Milano, 1989.

[Gue94] O. v. Guericke. *The New (So-Called) Magdeburg Experiments of Otto von Guericke*, volume 137 of *Archives Internationales d'Histoire des Idées*. Kluwer, Dordrecht, 1994. Translation and preface by M. G. F. Ames.

[Gui05] A. P. Guimarães. *From Lodestone to Supermagnets*. Wiley, Berlin, 2005.

[Haua] F. Hauksbee. An account of an experiment made before the Royal Society at Gresham-College, touching the extraordinary elistricity of glass, produceable on a smart attrition of it; with a continuation of experiments on the same subject, and other phenomena. *Philosophical Transactions*, 25:2327–2335, 1706–1707.

[Haub] F. Hauksbee. An account of the repetition of an experiment touching motion given bodies included in a glass, by the approach of a finger near its outside: With other experiments on the effluvia of glass. *Philosophical Transactions*, 26:82–86, 1708–1709.

[Hau09] F. Hauksbee. *Physico-Mechanical Experiments on Various Subjects, containing an Account of Several Surprising Phenomena touching Light and Electricity, producible on the Attrition of Bodies. With many other Remarkable Appearances, not before observ'd.* R. Brugis, London, 1709.

[Hea67] N. H. d. V. Heathcote. The early meaning of electricity: Some Pseudodoxia Epidemica—I. *Annals of Science*, 23:261–275, 1967.

[Hei79] J. L. Heilbron. *Electricity in the 17th & 18th Centuries.* University of California Press, Berkeley, 1979.

[Hei81a] J. L. Heilbron. Aepinus, Franz Ulrich Theodosius. In C. C. Gillispie, editor, *Dictionary of Scientific Biography,* Vol. 1, pages 66–68. Charles Scribner's Sons, New York, 1981.

[Hei81b] J. L. Heilbron. Dufay (Du Fay), Charles-François de Cisternay. In C. C. Gillispie, editor, *Dictionary of Scientific Biography,* Vol. 4, pages 214–217. Charles Scribner's Sons, New York, 1981.

[Hei81c] J. L. Heilbron. Gray, Stephen. In C. C. Gillispie, editor, *Dictionary of Scientific Biography,* Vol. 5, pages 515–517. Charles Scribner's Sons, New York, 1981.

[Hei81d] J. L. Heilbron. Hauksbee, Francis. In C. C. Gillispie, editor, *Dictionary of Scientific Biography,* Vol. 6, pages 169–175. Charles Scribner's Sons, New York, 1981.

[Hei81e] J. L. Heilbron. Nollet, Jean-Antoine. In C. C. Gillispie, editor, *Dictionary of Scientific Biography,* Vol. 10, pages 145–148. Charles Scribner's Sons, New York, 1981.

[Hei82] J. L. Heilbron. *Elements of Early Modern Physics.* University of California Press, Berkeley, 1982.

[Hei99] J. L. Heilbron. *Electricity in the 17th and 18th Centuries — A Study in Early Modern Physics.* Dover, New York, 1999.

[Hem80] l'abbé Hemmer. Sur l'électricité des métaux. *Journal de Physique*, 16:50–52, 1780.

[Hom67] R. W. Home. Francis Hauksbee's theory of electricity. *Archive for the History of Exact Sciences*, 4:203–217, 1967. Reprinted in R. W. Home, *Electricity and Experimental Physics in 18th-Century Europe* (Variorum, Hampshire, 1992), Chapter III.

[Hom81] R. W. Home. *The Effluvial Theory of Electricity.* Arno Press, New York, 1981.

[Jac99] J. D. Jackson. *Classical Electrodynamics.* John Wiley & Sons, New York, third edition, 1999.

[Kel81] S. Kelly. Gilbert, William. In C. C. Gillispie, editor, *Dictionary of Scientific Biography,* Vol. 5, pages 396–401. Charles Scribner's Sons, New York, 1981.

[Kra81] F. Krafft. Guericke, (Gericke), Otto von. In C. C. Gillispie, editor, *Dictionary of Scientific Biography,* Vol. 5, pages 574–576. Charles Scribner's Sons, New York, 1981.

[Lae91] D. Laertius. *Lives of Eminent Philosophers*, volume 1. Harvard University Press, Cambridge, 1991. Translated by R. D. Hicks.

[Llo80] J. T. Lloyd. Lord Kelvin demonstrated. *The Physics Teacher*, 18:16–24, 1980.

[Max81] J. C. Maxwell. *An Elementary Treatise on Electricity.* Clarendon Press, Oxford, 1881.

[Med02] A. Medeiros. As origens históricas do eletroscópio. *Revista Brasileira de Ensino de Física*, 24:353–361, 2002.

[Mel98] M. A. Melehy. Thermal momentum in thermodynamics, part 2. Interfacial electrification: a new consequence of the first and second laws. *Physics Essays*, 11:430–433, 1998.

[Net] L. F. Netto. Feira de ciências. Available at: www.feiradeciencias.com.br.

[New52a] I. Newton. *Mathematical Principles of Natural Philosophy*, volume 34, pp. 1-372 of *Great Books of the Western World.* Encyclopaedia Britannica, Chicago, 1952. Translated by A. Motte and revised by F. Cajori.

[New52b] I. Newton. *Optics*, volume 34, pp. 377-544 of *Great Books of the Western World.* Encyclopaedia Britannica, Chicago, 1952.

[Nol47] l'Abbé Nollet. Éclaircissemens sur plusieurs faits concernant l'électricité. *Mémoires de l'Académie royale des Sciences*, pages 102–131, 1747.

[Nol53] J. A. Nollet. *Essai sur l'Électricité des Corps.* Freres Guerin, Paris, 3rd edition, 1753.

[Nol67] J. A. Nollet. *Lettres sur l'électricité, dans lesquelles on trouvera les principaux phénomènes qui ont été découverts depuis 1760, avec des discussions sur les conséquences qu'on en peut tirer.* Durand, Neveu, Libraire, Paris, 1767. Troisieme partie.

[Ørs98a] H. C. Ørsted. New electro-magnetic experiments. In K. Jelved, A. D. Jackson, and O. Knudsen, editors, *Selected Scientific Works of Hans Christian Ørsted*, pages 421–424. Princeton University Press, Princeton, 1998. Paper originally published in German in 1820.

[Ørs98b] H. C. Ørsted. Observations on electro-magnetism. In K. Jelved, A. D. Jackson, and O. Knudsen, editors, *Selected Scientific Works of Hans Christian Ørsted*, pages 430–445. Princeton University Press, Princeton, 1998. Reprinted from Thomson's *Annals of Philosophy*, Vol. 2, pp. 321-337 (1821).

[Pla52a] Plato. *Protagoras.* In *Great Books of the Western World*, R. M. Hutchins, Editor in Chief, Vol. 7, pages 38–64, Chicago, 1952. Encyclopaedia Britannica. Translated by B. Jowett.

[Pla52b] Plato. *Timaeus.* In *Great Books of the Western World*, R. M. Hutchins, Editor in Chief, Vol. 7, pages 442–477, Chicago, 1952. Encyclopaedia Britannica. Translated by B. Jowett.

[Pri72] J. Priestley. An account of a new electrometer, contrived by Mr. William Henly, and of several electrical experiments made by him, in a letter from Dr. Priestley, F.R.S. to Dr. Franklin, F.R.S. *Philosophical Transactions*, 62:359–364, 1772.

[Pri66] J. Priestley. *The History and Present State of Electricity*, volume 1. Johnson Reprint Corporation, New York, 1966. The Sources of Science, Number 18. Reprinted from the third edition, London, 1755.

[Que] A. C. M. d. Queiroz. Electrostatic machines. Available at: www.coe.ufrj.br/~acmq/.

[RR53] D. Roller and D. H. D. Roller. The prenatal history of electrical science. *American Journal of Physics*, 21:343–356, 1953.

[RR57] D. Roller and D. H. D. Roller. The Development of the Concept of Electric Charge. In J. B. Conant, editor, *Harvard Case Studies in Experimental Science*, chapter 8, pages 541–639. Harvard University Press, Cambridge, 1957.

[Sas02] W. M. Saslow. *Electricity, Magnetism, and Light.* Academic Press, New York, 2002.

[The56] Theophrastus. *Theophrastus on Stones.* The Ohio State University, Columbus, 1956. Introduction, Greek text, English translation, and commentary by E. R. Caley and J. F. C. Richards.

[Tho] W. Thomson. On a self-acting apparatus for multiplying and maintaining electric charges, with applications to illustrate the voltaic theory. *Proceedings of the Royal Society of London*, 16:67–72, 1867–1868.

[Wal36] W. C. Walker. The detection and estimation of electric charges in the eighteenth century. *Annals of Science*, 1:66–100, 1936.

[Wil59] B. Wilson. Experiments on the tourmalin. *Philosophical Transactions*, 51:308–339, 1759.

[Zan81] B. Zanobio. Fracastoro, Girolamo. In C. C. Gillispie, editor, *Dictionary of Scientific Biography*, Vol. 5, pages 104–107. Charles Scribner's Sons, New York, 1981.

Made in the USA
Coppell, TX
04 March 2024

29709743R10148